Physics of Nonlinear Waves

Synthesis Lectures on Wave Phenomena in the Physical Sciences

Editor
Sanichiro Yoshida, *Southeastern Louisiana University*

Physics of Nonlinear Waves
Mitsuhiro Tanaka
2019

Gravitational Waves from Coalescing Binaries
Stanislav Babak
2019

Physics of Nonlinear Waves

Mitsuhiro Tanaka

ISBN: 978-3-031-01483-3   paperback
ISBN: 978-3-031-02611-9   ebook
ISBN: 978-3-031-00355-4   hardcover

DOI 10.1007/978-3-031-02611-9

A Publication in the Springer series
SYNTHESIS LECTURES ON WAVE PHENOMENA IN THE PHYSICAL SCIENCES

Lecture #2
Series Editor: Sanichiro Yoshida, *Southeastern Louisiana University*
ISSN pending.

Physics of Nonlinear Waves

Mitsuhiro Tanaka
Gifu University, Japan

SYNTHESIS LECTURES ON WAVE PHENOMENA IN THE PHYSICAL
SCIENCES #2

ABSTRACT

This is an introductory book about nonlinear waves. It focuses on two properties that various different wave phenomena have in common, the "nonlinearity" and "dispersion", and explains them in a style that is easy to understand for first-time students.

Both of these properties have important effects on wave phenomena. Nonlinearity, for example, makes the wave lean forward and leads to wave breaking, or enables waves with different wavenumber and frequency to interact with each other and exchange their energies.

Dispersion, for example, sorts irregular waves containing various wavelengths into gentler wavetrains with almost uniform wavelengths as they propagate, or cause a difference between the propagation speeds of the wave waveform and the wave energy.

Many phenomena are introduced and explained using water waves as an example, but this is just a tool to make it easier to draw physical images. Most of the phenomena introduced in this book are common to all nonlinear and dispersive waves.

This book focuses on understanding the physical aspects of wave phenomena, and requires very little mathematical knowledge. The necessary minimum knowledges about Fourier analysis, perturbation method, dimensional analysis, the governing equations of water waves, etc. are provided in the text and appendices, so even second- or third-year undergraduate students will be able to fully understand the contents of the book and enjoy the fan of nonlinear wave phenomena without relying on other books.

KEYWORDS

nonlinear wave, dispersive wave, water wave, soliton, modulated wavetrain, wave-wave interaction, wave turbulence

Contents

Preface

This is an introductory book about nonlinear waves. We humans are living surrounded by various wave phenomena. Sounds of people talking and birds singing are the sound waves transmitted through the air. The energy from the sun and TV programs from broadcasting stations are delivered by electromagnetic waves. If a strong wind blows over the sea surface, water waves are generated and travel thousands of kilometers as swells. The wave phenomena play important roles from our daily life to longer-term global environmental changes.

This book focuses on two properties that many of such diverse wave phenomena have in common, the "nonlinearity" and "dispersion", and explains them in a style that is easy to understand for the first-time students.

"Linear" is a term that means direct proportion, and thus "nonlinear" literally means out of proportion. For example, we learn in high school physics that the force to pull a spring and the extension of the spring are proportional, i.e., the familiar Hooke's law. This may be correct while the force is weak. However, it is also a well-known fact that the relation between the two deviates from direct proportion when the force and the resultant extension of the spring get larger. The same is true for the familiar Ohm's law $V = RI$ of electricity. This may also be correct while the electric current I is weak. However, if the current becomes larger, the temperature of the resister will rise due to the generated Joule heat, and as a result, the resistance R would increase. If this effect is included, R becomes a function of I, so the voltage V can no longer be in direct proportion to I. Thus, "nonlinearity" is ubiquitous and exists everywhere in our surroundings. With respect to wave phenomena, the nonlinearity plays various important roles that can not be captured within the linear framework, such as deforming the waveform with wave propagation, enabling energy exchange between waves with different frequencies, and so on.

Another keyword that is important throughout this book is "dispersion" of waves. Dispersion is the property that the wave travels at different speeds depending on the wavelength or frequency. Sound waves and light (electromagnetic waves) do not have dispersion. When asked, "What is the speed of light?", most people would immediately answer "300,000 km/s" or "seven and a half lap of earth in one second". Also, if asked "What is the speed of sound?", many people would answer "about 340 m/s". In these cases, any conditions such as "light of what color" or "sound of what Hz" are not appended. This is because the speed of these waves do not depend on frequency or wavelength.

Then how about if you are asked "What is the speed of water waves?" You can see the waves on the water surface on a daily basis, in the bathtub, on the river, etc. In the sense that it is visible, the water wave is more familiar than light or sound. However, most people do not know how fast its speed is. In fact, water wave is a dispersive wave, and its speed depends largely on wavelength. While the tsunami which travels through the Pacific Ocean is as fast as 200 m/s (over 700 km/h), the waves in the bathtub are less than 1 m/s. Waves that do not have dispersion like sound waves and electromagnetic waves are rather exceptional, and many waves around us have dispersion like water waves. This dispersion also brings about complex and interesting properties to wave phenomenon, as introduced in Chapter 3 as well as in subsequent chapters. Richard Feynman takes up the water wave as an example of wave phenomena following light and sound waves in his famous physics course.[1] There, with the dispersive nature of water wave in mind, he states: "water waves are the worst possible example, because they are in no respects like sound and lights; they have all the complications that waves can have."

In this book, I will introduce in an easy-to-understand manner various intriguing wave phenomena that "nonlinearity" and "dispersion" produce and the physical mechanisms underlying them. In many cases, they are described in the context of water waves. However, it should be noted that the water wave is used only to facilitate understanding of the mechanism behind each phenomenon and building concrete physical images, and most of the phenomena covered in this book are common in many different kinds of waves that have both nonlinearity and dispersion.

There are two techniques that are often used throughout the book. It is "dimensional analysis" and "perturbation method." These are highly versatile and important analysis methods used in a wide range of research fields, not limited to wave studies. For those readers who have not learned these subjects, perturbation method is described in somewhat detail in Chapter 4, and the basics of dimensional analysis is described in Appendix E. I have also prepared Appendix D where a brief review on minimum knowledge of fluid mechanics and the derivation method of the governing equations for water surface waves are given for the readers who are not familiar with fluid mechanics. With the help of these chapters and appendices, I believe that the reader can fully understand the contents of this book and enjoy the fan of nonlinear wave phenomena without relying on other books if he or she has elementary knowledge of partial differentiation and the Fourier analysis, and a little willingness to learn.

Mitsuhiro Tanaka
Gifu, Japan, November 2019

---

[1]Feynman, R, Leighton, R. and Sands, M.: *The Feynman Lectures on Physics*, (Addison-Wesley, Basic Books, 1964)

Acknowledgments

I would like to express my warmest thanks to my supervisor, Takuji Kawahara, who unfortunately passed away this year, from whom I learned nonlinear waves firsthand as a Ph.D. student of Kyoto University. I am grateful to Prof. Hidekazu Tsuji of Kyushu University, whose friendship with Prof. Sanichiro Yoshida, the editor of the series, has made this publication possible. Finally, I thank my wife Mayumi and two daughters, Haru and Natsuki, for their encouragement at every stage of preparation of this book.

Mitsuhiro Tanaka
November 2019

CHAPTER 1

The Simplest Nonlinear Wave Equation

In this chapter, we will explain about what is "wave" and what is the simplest equation to describe it, how to solve the equation to predict the waveform at later times, what kind of effects the "nonlinearity" produces, and so on.

1.1 THE SIMPLEST WAVE EQUATION

What would you answer if you are asked, "What is a wave?" Of course, there may be various ways to answer. In this book, we will call the phenomenon a "wave" in which any change of some observable quantity that occurs at a certain point is transmitted to a spatially separated place with time. According to this definition, the "wave" that travels through the audience of a stadium that you see in heated games of baseball or football is literally a wave phenomenon.

If we think of wave in this sense, the most ideal wave would be that a signal (waveform) travels at a constant speed c without changing its shape. Let $f(x,t)$ denote the value of a physical quantity (for example, the head height of the audience in the case of the "waves" going around the stadium) at the location x and the time t. Then, in such an ideal wave, the waveform $f(x,t)$ at an arbitrary later time t is simply given by the translation of the initial waveform $F(x)$ by ct, that is, $f(x,t) = F(x - ct)$, as shown in Fig. 1.1.

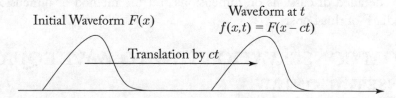

Initial Waveform $F(x)$

Waveform at t
$f(x,t) = F(x - ct)$

Translation by ct

Figure 1.1: Simple tanslation without changing the waveform.

What would be the simplest mathematical formula that expresses this simple translation at a constant speed without a change of form? If $x - ct$ is written as ξ, $f(x,t) = F(\xi)$, that is, $f(x,t)$ is a function of ξ only, and depends on x and t only through ξ. Then according to the chain rule of partial differentiation,

$$\frac{\partial f}{\partial t} = \frac{dF}{d\xi}\frac{\partial \xi}{\partial t} = -c\frac{dF}{d\xi}, \quad \frac{\partial f}{\partial x} = \frac{dF}{d\xi}\frac{\partial \xi}{\partial x} = \frac{dF}{d\xi}. \tag{1.1}$$

Therefore, when the function $f(x,t)$ translates at a speed c, $f(x,t)$ satisfies the Partial Differential Equation (written as PDE for short)

$$\frac{\partial f}{\partial t} + c\frac{\partial f}{\partial x} = 0, \tag{1.2}$$

no matter what the shape of the waveform $F(\xi)$ is. Thus, Eq. (1.2) is the simplest PDE that expresses the typical wave phenomenon that a signal travels in the x direction with a velocity c without changing the waveform. If you see the operator $\frac{\partial}{\partial t} + c\frac{\partial}{\partial x}$, you can automatically think of it as an operator that expresses the simple translation with a speed c.

It should be noted that since a wave is a phenomenon in which a change in a physical quantity f that occurs at a certain place x is transmitted to another place with time t, f must be a multi-variable function, denoted as $f(x,t)$.[1] In addition, what is important in wave phenomenon is the change (or derivative) of $f(x,t)$ that accompanies the arrival and passage of wave, and thus, if the wave phenomenon is expressed in terms of mathematical expressions, the differential equation for a multivariable function, or PDE, inevitably appears.

By the way, most of the physical quantities around us have "dimensions," such as the dimension of "length" and the dimension of "mass." And, as a question such as "Which is heavier, 1 m or 1 kg?" is nonsense, we cannot compare, add, or subtract quantities of different dimensions. In the case of Eq. (1.2), denoting the dimensions of time and length by T and L, respectively, and the dimensions of c and f by $[c]$ and $[f]$, respectively, then the dimensions of the first and the second terms on left side are $[f]/T$ and $[c][f]/L$, respectively. Then in order for these two terms to have the same dimension, the dimension $[c]$ must be L/T. That is, when there is an operator $\frac{\partial}{\partial t} + c\frac{\partial}{\partial x}$, the coefficient c before the spatial derivative $\frac{\partial}{\partial x}$ is always a quantity with the dimension of "velocity."

For more detailed discussions on dimensions and the method of dimensional analysis, refer to Appendix F of this book.

1.2   FROM CONSERVATION LAW TO WAVE EQUATION

1.2.1   CONSERVATION LAW

As mentioned in the Preface, there are a lot of wave phenomena around us. This implies that there are many phenomena that are approximately expressed by an equation like (1.2) occurring everywhere. One of the typical routes that an equation like (1.2) appears is the combination of a conservation law and a constitutive equation (or an equation of state). The laws that govern

---

[1]When considering waves in the more general 3D xyz space, we need to treat functions of four independent variables like $f(x,y,z,t)$.

almost all motions and phenomena in the world, from the atomic scale to the scale of the universe, are conservation laws, such as conservation laws of mass, momentum, energy, etc. In the following, we will explain the concept of the conservation law for continuous media, using the flow of electricity in a wire as shown in Fig. 1.2 as an example.

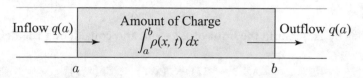

Figure 1.2: Flow of charge in a wire.

The charge is conserved. In other words, it does not suddenly appear from nowhere or disappear in the empty sky. Focusing on a particular section of the wire, the amount of charge present there may not be constant but increase or decrease with time. But it is simply because the charge flows into or out of the section through the ends of the section. When we consider the equation that the charge distribution along the wire should follow, the important physical quantities to consider are the density of charge $\rho(x,t)[\text{C/m}]$ and the flux of charge $q(x,t)[\text{C/s}]$, where m is meter, s is second, and C is coulomb. The flux of charge $q(x,t)$ represents the amount of charge that passes the point of interest x per 1 s in the positive x direction.[2]

Now, focusing on an arbitrary section $a \le x \le b$ of a wire, consider the change in the amount of charge contained in this section during a very short time Δt. The amount of charge contained in this section at times t and $t + \Delta t$ are given by $\int_a^b \rho(x,t)\,dx$ and $\int_a^b \rho(x,t+\Delta t)\,dx$, respectively, thus the increment during Δt is given by

$$\int_a^b [\rho(x,t+\Delta t) - \rho(x,t)]\,dx. \tag{1.3}$$

On the other hand, the amount of charge flowing into this part through $x = a$ and that flowing out at $x = b$ are $q(a,t)\Delta t$ and $q(b,t)\Delta t$, respectively, so the net inflow during Δt is given by

$$[q(a,t) - q(b,t)]\,\Delta t. \tag{1.4}$$

Here, we assume that Δt is a very short time interval. Otherwise, the inflow over Δt must be written as an integral as $\int_t^{t+\Delta t} q(a,t')\,dt'$. That is, the meaning that "Δt is very short" means that it is very short compared to the time interval in which q and ρ change significantly. When using words such as "small," "short," etc., it is always necessary to be aware of what they are compared with.

---

[2]The flux of charge is nothing but the current. The definition of 1 ampere [A] is 1[C/s].

The conservation law of charge requires that the increment of charge (1.3) is equal to the net inflow (1.4), that is:

$$\int_a^b [\rho(x, t + \Delta t) - \rho(x, t)] \, dx = [q(a, t) - q(b, t)] \, \Delta t. \tag{1.5}$$

Dividing both sides by Δt, taking the limit of $\Delta t \to 0$, and considering the definition of partial derivative

$$\lim_{\Delta t \to 0} \frac{\rho(x, t + \Delta t) - \rho(x, t)}{\Delta t} = \frac{\partial \rho(x, t)}{\partial t}, \tag{1.6}$$

yields

$$\int_a^b \frac{\partial \rho}{\partial t} \, dx = q(a, t) - q(b, t). \tag{1.7}$$

If we write the right side by an integral as

$$q(a, t) - q(b, t) = -\int_a^b \frac{\partial q}{\partial x} \, dx, \tag{1.8}$$

and put it on the left side, we get

$$\int_a^b \left(\frac{\partial \rho}{\partial t} + \frac{\partial q}{\partial x} \right) dx = 0. \tag{1.9}$$

Since the interval of interest $[a, b]$ is arbitrary, and this integral must always be 0 regardless of a and b. From this, the equation

$$\frac{\partial \rho(x, t)}{\partial t} + \frac{\partial q(x, t)}{\partial x} = 0 \tag{1.10}$$

should hold everywhere.

In the above, we talked about the flow of charge in a wire as an example. However, as understood from the derivation process, if a certain physical quantity (with unit $\odot$) is conserved, the relationship of the equation (1.10) always holds between its line density $\rho(x, t)[\odot/\mathrm{m}]$ and its flux $q(x, t)[\odot/\mathrm{s}]$. In the more general 3D case, the 3D conservational PDE of the form

$$\frac{\partial \rho}{\partial t} + \nabla \cdot q = 0, \tag{1.11}$$

holds between the density $\rho(x, t) \, [\odot/\mathrm{m}^3]$ and the flux vector $q(x, t) \, [\odot/\mathrm{m}^2/s]$. Here, $\nabla \cdot q$ is a quantity called "the divergence of vector q." (For more detail, see Appendix A.)

1.2.2 FROM CONSERVATION LAW TO WAVE EQUATION

The conservation equation (1.10) contains two unknowns: the density $\rho(x, t)$ and the flux $q(x, t)$. In this situation, the number of equation is insufficient, and it is impossible to trace the temporal evolution of $\rho(x, t)$ even if the initial distribution is specified.[3] There are two typical ways of solving this problem as follows:

1. deriving a new equation $\dfrac{\partial q}{\partial t} = g(\rho, q)$ describing the time evolution of $q(x, t)$ from some physical law such as other conservation law, and construct a "closed" initial value problem; and

2. deriving an algebraic relation $q = q(\rho)$ that relates ρ and q from some rule or observation result. Such relation is called the equation of state or the constitutive law.[4]

In the latter case, substituting the equation of state $q = q(\rho)$ in (1.10) immediately gives a wave equation like (1.2) as follows:

$$(1.10) \quad \longrightarrow \quad \frac{\partial \rho}{\partial t} + c(\rho)\frac{\partial \rho}{\partial x} = 0, \quad \text{where} \quad c(\rho) \equiv \frac{dq(\rho)}{d\rho}. \tag{1.12}$$

As a simple example of such a problem that can be reduced to a wave equation (1.2) by combining the conservation law and the equation of state, let us introduce a simple model of traffic flow below.

1.2.3 SIMPLE MODEL OF TRAFFIC FLOW

For a one-lane road with no merging or branching as shown in Fig. 1.3, consider how the density of cars changes with space and time. The density and flux of cars are denoted by $\rho(x, t)$ [cars/km] and $q(x, t)$[cars/h], respectively.[5]

If there are no branches of roads, the number of cars on the road is a conserved amount, and the conservation equation (1.10) holds between ρ and q. Also, if we write the speed at which the car runs as $v(x, t)$ [km/h], $q = \rho v$ holds, then by substituting this in (1.10) gives

$$\frac{\partial \rho}{\partial t} + \frac{\partial (\rho v)}{\partial x} = 0. \tag{1.13}$$

---

[3]This situation is often described as "The problem is not closed."

[4]In order for such an algebraic relation that does not include a time derivative to hold, q must have the ability to instantly follow ρ no matter how fast ρ changes. Strictly speaking, such a relationship holds only in an equilibrium state where ρ and q do not change with time, or in a quasi-static process that changes very slowly.

[5]Here, imagine that you are not looking at a road from a short distance where each car can be identified, but looking at the road from far above where you can no longer identify each car but can only see the "pattern of crowdedness" along the road over a much longer range than the length of a car. This point of view is similar to that which we employ in fluid mechanics. When we deal with air by fluid mechanics, for example, the density and flow velocity of air are treated as if they are continuous functions of x and t, in spite of the fact that, microscopically, air consists of many nitrogen and oxygen molecules flying around in vacuum, hence are of quite discrete nature. The concept of "continuum" lies in the background of such treatment of gases and liquids in fluid mechanics.

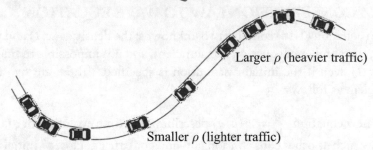

Figure 1.3: Traffic on a one-lane road.

Equation (1.13) contains two unknowns, ρ and v, and the problem is not closed. The simplest model to make the problem closed is to assume that the velocity v is determined by the density ρ, and assume some equation of state $v = v(\rho)$. It would be reasonable to assume that the function $v(\rho)$ reaches a maximum value v_0 as $\rho \to 0$, becomes 0 at the density ρ_{jam} corresponding to a perfect traffic jam where cars cannot move at all, and is a monotonically decreasing function of ρ between $\rho = 0$ and ρ_{jam}. Figure 1.4 shows schematically such a relation between ρ and v, while Fig. 1.5 shows the corresponding relation between ρ and $q\ (= \rho v)$.

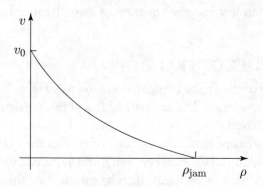

Figure 1.4: Typical relationship between ρ and v.

Figure1.5 shows that q takes the maximum at a certain density ρ_{m}.

As shown in (1.12), the velocity c with which the density of cars is transmitted is given by $c = dq(\rho)/d\rho$, and its outline becomes something like that shown in Fig. 1.6.

The figure shows that, for an observer stationary with respect to the road, $c < 0$ and the change in ρ propagates toward the rear of the road when $\rho > \rho_{\text{m}}$, while $c > 0$ and the change in ρ conversely propagates toward the front of the road when $\rho < \rho_{\text{m}}$.

From the relation

$$c(\rho) = \frac{dq(\rho)}{d\rho} = \frac{d(\rho v(\rho))}{d\rho} = v(\rho) + \rho \frac{dv(\rho)}{d\rho}, \qquad (1.14)$$

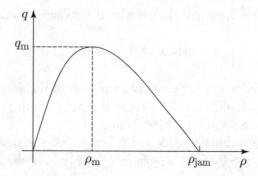

Figure 1.5: Typical relationship between ρ and q.

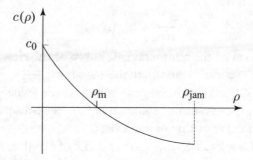

Figure 1.6: Typical relationship between ρ and c.

and $\frac{dv(\rho)}{d\rho} < 0$, $c < v$ always holds. This means that the speed v at which the car runs is always faster than the speed c at which the change in ρ propagates, so that the car catches up with the "wave" of change in ρ from behind. From the car driver's point of view, this means that the information on changes (such as the road is getting more crowded or more vacant) always comes from ahead of him/her. This seems to be consistent with everyday experience.

1.3   METHOD OF CHARACTERISTICS

If the propagation velocity c is a constant ($= c_0$), the wave equation is given by

$$\frac{\partial \rho}{\partial t} + c_0 \frac{\partial \rho}{\partial x} = 0. \tag{1.15}$$

This equation is linear with respect to the unknown function $\rho(x, t)$. Here, as described at the beginning of this chapter, any initial waveform translates at speed c_0. If the initial waveform is given by $\rho = \rho_0(x)$, then the waveform at time t is given by $\rho(x, t) = \rho_0(x - c_0 t)$. However, when derived in the way explained in (1.12), the resultant wave equation becomes

$$\frac{\partial \rho}{\partial t} + c(\rho)\frac{\partial \rho}{\partial x} = 0, \tag{1.16}$$

in which the propagation velocity c is generally a function of the dependent variable ρ. In this case, the wave equation is a nonlinear PDE.[6] Here, it is not so simple to find $\rho(x,t)$ at any time t. Let us consider the solution of this problem below.

The density $\rho(x,t)$ is a function of x and t, and its value is defined at each point of the upper-half $-\infty < x < \infty$, $t \geq 0$ of the xt-plane. We only have the freedom to give the initial distribution $\rho(x,0)$ on the x-axis (i.e., $t = 0$), and $\rho(x,t)$ at $t > 0$ is automatically determined from the initial condition and the governing equation (1.16). A curve $C : x = X(t)$ on the upper-half xt plane such that

$$\frac{dX}{dt} = c[\rho(X,t)] \tag{1.17}$$

holds at each point on it is called the **characteristic curve** or the **characteristics** of (1.16). That is, that C is a characteristic curve means that the reciprocal $\frac{dX(t)}{dt}$ of the slope at each point $(X(t),t)$ on C is equal to the value of c corresponding to the value of ρ at that point. Before solving the problem, such a curve cannot be known in advance because $\rho(x,t)$ is not known yet. But it is certain that such a curve of this nature exists.

From (1.17), along C, a small changes Δt and Δx of t and x, respectively, are related by $\Delta x = c(\rho)\Delta t$. So the small change $\Delta \rho$ of $\rho(x,t)$ along C that occurs during a small time Δt is given by

$$\Delta \rho = \rho(X + \Delta x, t + \Delta t) - \rho(X,t) = \frac{\partial \rho}{\partial t}\Delta t + \frac{\partial \rho}{\partial x}\Delta x$$
$$= \left[\frac{\partial \rho}{\partial t} + c(\rho)\frac{\partial \rho}{\partial x}\right]\Delta t = 0, \tag{1.18}$$

where (1.16) was used for the last part. From this, it can be seen that $\rho(x,t)$ does not change along the characteristics C, that is, the characteristics C is a curve carrying a constant value of ρ. Since the slope of C is given by $1/c(\rho)$, if ρ is constant along C, then the slope of C is also constant, so C becomes a straight line.

Therefore, the initial value problem of the nonlinear wave equation (1.16) can be solved by the following procedure.

1. From the initial distribution $\rho_0(x)$ of ρ, find the initial distribution of c by $c_0(x) = c[\rho_0(x)]$.

---

[6]When we say "linear" and "nonlinear" in the context of differential equations, we do not care if the dependent variable is differentiated. Also, it does not matter if the independent variable is multiplied. For example, $\rho\rho_x$ is a quadratic term and $x^5\rho_x$ is a linear term. For nonlinear differential equations, the powerful method of "superposition of solutions" that holds true for linear differential equations does not hold, which makes solving them much more difficult.

2. Passing an arbitrary point $(x, t) = (\xi, 0)$ on the x axis, draw a straight line with a slope $1/c_0(\xi)$, i.e., $x = \xi + c_0(\xi)t$. This is a characteristics on which ρ takes a constant value $\rho_0(\xi)$.

3. Then, the waveform $\rho(x, t)$ at time t is given by a parametric representation using the starting point ξ of the characteristics as a parameter as follows:

$$\rho(\xi) = \rho_0(\xi), \qquad x(\xi) = \xi + c_0(\xi)t. \tag{1.19}$$

This solution procedure is called the **method of characteristics**.

EXAMPLE 1: THE METHOD OF CHARACTERISTICS

By using the method of characteristics, draw the waveform of $\rho(x, t)$ at $t = 5$ of the solution of the initial value problem of the nonlinear wave equation:

$$\rho_t + \rho\rho_x = 0, \quad \rho(x, 0) = \begin{cases} 0 & (x \leqq 0), \\ e^{-1/x} & (x > 0). \end{cases} \tag{1.20}$$

[Answer]

For example, when using Microsoft Excel, the procedure may be as follows.

1. Name the first column of the spreadsheet ξ, and enter numerical values at interval of 0.1 from -5 to 10. (Select the range and interval here appropriately.)

2. Name the second column ρ and if the first column $\xi \leq 0$, put $\rho = 0$, and if $\xi > 0$, put $\rho = e^{-1/\xi}$.

3. Name the third column c and enter the speed of the characteristics corresponding to each ρ. In the present case, $c = \rho$, so put the same value as in the second column.

4. Name the fourth column x. Here, enter the x-coordinate at which the characteristics starting from each ξ in the first column reaches at $t = 5$ at a speed of $c(\xi)$, that is, enter the number with the formula $x = \xi + c(\xi) \times 5$.

If this preparation is made, the initial waveform can be obtained by drawing the graph of the second column ρ with the first column ξ as the horizontal axis, and you get the waveform at $t = 5$ if you draw the second column ρ with the fourth column x as the horizontal axis. Figure 1.7 shows a part of the spreadsheet and the waveform obtained by this procedure.

In the linear wave equation (1.15), the waveform only translates and remains the same forever. It should be noted that the change in waveform as we have seen here arises solely from the "nonlinearity," that is, the propagation velocity c depending on the dependent variable ρ. ♣

....

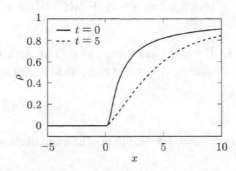

▲	A	B	C	D
1			time=	5
2	xi	rho	c	x
52	−0.1	0.000	0.000	−0.100
53	0.0	0.000	0.000	0.000
54	0.1	0.000	0.000	0.100
55	0.2	0.007	0.007	0.234
56	0.3	0.036	0.036	0.478
57	0.4	0.082	0.082	0.810
58	0.5	0.135	0.135	1.177
59	0.6	0.189	0.189	1.544
60	0.7	0.240	0.240	1.898
61	0.8	0.287	0.287	2.233
62	0.9	0.329	0.329	2.546
63	1.0	0.368	0.368	2.839
64	1.1	0.403	0.403	3.115
65	1.2	0.435	0.435	3.373
66	1.3	0.463	0.463	3.617
67	1.4	0.490	0.490	3.848
68	1.5	0.513	0.513	4.067
69	1.6	0.535	0.535	4.276

Figure 1.7: The solution of Example 1.

By applying the method of characteristics shown above, let us consider the time change of the queue of cars that occurs when the signal changes from red to green. Assuming that the location of the signal is $x = 0$, the initial distribution of car density ρ is given by

$$\rho_0(x) = \begin{cases} 0 & (x > 0) \\ \rho_{\text{jam}} & (x < 0). \end{cases} \tag{1.21}$$

Note that when $x = 0$, ρ is interpreted as taking all values between 0 and ρ_{jam}.

The pattern of the characteristics becomes as shown in Fig. 1.8. The head of the line of the cars, that is, the characteristics C_+ separating the region with $\rho = 0$ from that with $\rho \neq 0$ travels at a constant velocity of c for $\rho = 0$, i.e., $c(0)$. On the other hand, the characteristics C_- separating the region in which the cars has already started moving and the part in which the cars cannot move yet travels at a constant velocity c for ρ_{jam}, i.e., $c(\rho_{\text{jam}})$. Since $c(\rho_{\text{jam}}) < 0$, C_- propagates backward. Between C_+ and C_-, there are infinite number of characteristics corresponding to $0 < \rho < \rho_{\text{jam}}$ which are transmitted at each speed $c(\rho)$ from the origin, forming a fan-shaped area as shown in Fig. 1.8. It can be seen that the corresponding waveform of $\rho(x, t)$ changes, as shown in Fig. 1.9.[7]

A car at $x = -D$ in the waiting queue cannot move until the characteristics C_- arrives at it, even if the signal has changed to green. It can start moving $D/|c(\rho_{\text{jam}})|$ after the signal turns green. This is an inevitable consequence derived from a mathematical model of the traffic flow, and it cannot be helped even if you get irritated and scream "Why can't I move even though the signal has changed green?"

---

[7]It can be shown that the waveform of $\rho(x, t)$ in the fan-shaped area is given by the graph of $c(\rho)$ shown in Fig. 1.6 after rotating the figure to make c the horizontal axis and ρ the vertical axis, and stretch the horizontal axis by t times from c to ct.

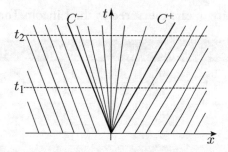

Figure 1.8: Characteristic curves when the signal turns green.

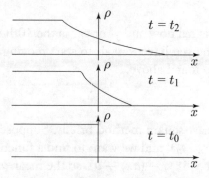

Figure 1.9: Waveform of ρ at the initial and later times.

It can also be seen from Fig. 1.8 that the characteristics of propagation velocity $c = 0$ is always at the signal position $x = 0$. Since $c = dq(\rho)/d\rho$, this is the characteristics that corresponds to ρ_{m}, which maximizes the flux q. This means that if the queue does not end, at the position of the signal, the state of maximum flux, i.e., the state where the road is used most effectively, continues to be realized automatically.

EXAMPLE 2: APPLICATION OF METHOD OF CHARACTERISTICS TO TRAFFIC FLOW

At the Lincoln Tunnel connecting Manhattan, New York, and New Jersey, Greenberg [1] actually measured in 1959 the density and speed of vehicles, and reported the results as shown in Table 1.1. In the table, the density and the speed are given in units of (cars/mile) and (mile/h), respectively, where 1 mile = 1.609 km.

1. Find the relationship $v(\rho) = a \log \rho + b$ between ρ and v that best fits the data given in the table using the least square method. Also, from the result, find the density ρ_{jam} when the cars are completely congested and $v = 0$.

Table 1.1: Density and velocity of cars measured at the Lincoln Tunnel

Density ρ	Velocity ν	Density ρ	Velocity ν	Density ρ	Velocity ν
34	32	88	17	108	11
44	28	94	16	129	10
53	25	94	15	132	9
60	23	96	14	139	8
74	20	102	13	160	7
82	19	112	12	165	6

2. Suppose that the signal is red now and your car is the 10th from the head of the waiting queue. Estimate how long you have to wait to start moving when the signal turns green, based on the relationship found above.

[Answer]

1. First, let us review the least-square method briefly. Suppose that there are n pairs of real numbers $(x_1, y_1), \cdots, (x_n, y_n)$, and we want to find a function $y = ax + b$ that approximates these. The error at x_i is $y_i - (ax_i + b)$, so the mean value of the squared errors per point , i.e., the mean squared error E is given by

$$E = \frac{1}{n} \sum_{i=1}^{n} \{y_i - (ax_i + b)\}^2 . \tag{1.22}$$

Here, if the errors at each point were simply summed up, the errors with opposite signs would cancel each other out and would not be reflected in E, so the errors are squared and then summed.

E is a function of a and b, and a and b that minimize E must satisfy

$$\frac{\partial E(a, b)}{\partial a} = 0, \quad \frac{\partial E(a, b)}{\partial b} = 0. \tag{1.23}$$

This condition gives a set of simultaneous linear equations for a and b as follows:

$$\left(\sum_{i=1}^{n} x_i^2 \right) a + \left(\sum_{i=1}^{n} x \right) b = \left(\sum_{i=1}^{n} x_i y_i \right), \tag{1.24a}$$

$$\left(\sum_{i=1}^{n} x_i \right) a + \left(\sum_{i=1}^{n} 1 \right) b = \left(\sum_{i=1}^{n} y_i \right), \tag{1.24b}$$

and by solving these, the coefficients a and b of the function $y = ax + b$ that best approximates (x_i, y_i) $(i = 1, \cdots n)$ in the sense that the mean square error E is minimized are obtained.

By applying this least-square method with $x = \log \rho$ and $y = v$, and converting the unit of length from miles to km, we obtain

$$v(\rho) = a \log \rho + b, \quad a \approx -27.3 \,[\text{km/h}], \quad b \approx 135.7 \,[\text{km/h}]. \tag{1.25}$$

Also by setting $v = 0$ here, $\rho_{\text{jam}} = \exp(-b/a) \approx 143 \,[\text{car/km}]$.

2. From the result of 1.

$$c(\rho) = \frac{d(\rho v)}{d\rho} = v + a \quad \longrightarrow \quad c(\rho_{\text{jam}}) = a \approx -27.3 \,[\text{km/h}]. \tag{1.26}$$

Then, the time τ that takes to start moving per one vehicle is given by

$$\tau = \frac{1/\rho_{\text{jam}}}{c(\rho_{\text{jam}})} = 2.56 \times 10^{-4} \,[\text{h/car}] \approx 0.92 \,[\text{s/car}]. \tag{1.27}$$

Thus, it is expected that it will take about 9 seconds for the 10th car to start moving.

♣

However, it should be noted that there is a serious problem with this estimation. In the simple model treated here, the traffic flow is formulated based on the viewpoint from far above where each vehicle cannot be seen. As a result of such a viewpoint, we are allowed to represent the original discrete system consisting of individual cars as a continuum which can be expressed by using continuous functions such as $\rho(x, t)$ and $v(x, t)$. If we remember this, discussing the movement of 10 cars or so, as in this example, seems to somewhat deviate from the range of validity of the formulation. In any case, you can check on your own how the above result is reasonable when waiting for a signal.

· ·

1.4   INTERSECTION OF CHARACTERISTICS AND OCCURRENCE OF MULTIVALUEDNESS

As we saw above, in the nonlinear wave equation (1.16), a point with a value of ρ travels along the characteristics at a speed of $c(\rho)$. If, in the initial waveform, ρ corresponding to fast c is behind ρ corresponding to slow c, then with the passage of time, the fast characteristics starting from behind catches up with the slow characteristics starting from the front, and at some point in time an intersection of characteristics occurs. In the case of traffic flow, c is a monotonically decreasing function of ρ, so this happens when there is a part with larger ρ (more crowded) in

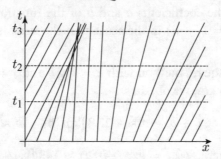

Figure 1.10: Intersection of characteristics.

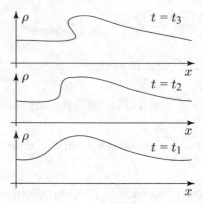

Figure 1.11: Steepening of waveform.

front of a part with smaller ρ (less crowded), i.e., a part with $\partial\rho/\partial x > 0$ in the initial waveform. (See Fig. 1.10).

Figure 1.11 shows the time change of the corresponding waveform of ρ. It can be seen that the waveform becomes steeper as the characteristics carrying different ρ approach with time, and the slope diverges at a certain time, and the waveform finally becomes a multivalued function after that time. In the context of traffic flow, ρ represents the density of cars (number of cars per km), and it is not physically permitted to have multiple values at one point in space-time. In other words, we cannot but say that the theory has broken when the multivaluedness occurs.

As shown in Fig. 1.11, in the waveform of car density, the part with $\frac{\partial\rho}{\partial x} > 0$ tends to steepen automatically with time. As shown by (1.14), the speed v the car actually runs is always faster than $c(\rho)$, so the person driving the car enters the waveform of $\rho(x,t)$ as shown in Fig. 1.11 from the left, passes through it and leaves to the right. Then, the fact that the part of the waveform of ρ where $\partial\rho/\partial x > 0$ has a tendency to steepen implies that, when a car encounters a crowded state it happens suddenly, on the other hand, when it leaves the crowded area the change occurs only slowly. This seems to be consistent with everyday experience again.

In the nonlinear wave equation (1.16), when the initial waveform $\rho_0(x) = \rho(x, 0)$ is specified, the time when the characteristics intersect for the first time and the multivaluedness occurs can be predicted as follows. Since the propagation velocity c of the characteristics carrying a certain value of ρ is determined by the value of ρ, given the initial distribution $\rho_0(x)$ of ρ, the initial distribution $c_0(x) = c[\rho_0(x)]$ is determined. Let us imagine how the characteristics of propagation velocity $c_0(x)$ leaving x initially catches up with the characteristics of propagation velocity $c_0(x + \Delta x)$ starting from $x + \Delta x$. Since $c_0(x + \Delta x) \approx c_0(x) + (dc_0/dx)\Delta x$ when $\Delta x \ll 1$, the speed difference Δc between them is given by $(dc_0/dx)\Delta x$. ($dc_0/dx < 0$ for an intersection of characteristics to occur.) Then the time required for the velocity difference Δc to cancel the initial spatial gap of Δx is $|\Delta x/\Delta c| = 1/|dc_0/dx|$. Therefore, the time t_b at which the intersection occurs for the first time at somewhere of waveform is given by

$$t_b = \min\left\{\frac{1}{|dc_0/dx|}\right\} = \frac{1}{\max\{|dc_0/dx|\}}, \tag{1.28}$$

where min{ }, max{ } mean the maximum and the minimum values over the whole initial waveform, respectively.

EXAMPLE 3: STEEPENING AND OCCURRENCE OF MULTIVALUEDNESS

For the initial value problem of nonlinear wave equation

$$\frac{\partial \rho}{\partial t} + \rho \frac{\partial \rho}{\partial x} = 0, \qquad \rho_0(x) = e^{-x^2}, \tag{1.29}$$

predict the time t_b at which the solution becomes multivalued for the first time.

[Answer]

In this problem, $c(\rho) = \rho$, so $c_0(x) = \rho_0(x) = e^{-x^2}$. Then, $-dc_0/dx = 2x\,e^{-x^2}$, which takes the maximum value $\sqrt{2/e}$ at $x = 1/\sqrt{2}$. Therefore, t_b is given by $\sqrt{e/2}$, and at this time those characteristics that start from the vicinity of $x = 1/\sqrt{2}$ at $t = 0$ are expected to intersect. Figure 1.12 shows the time evolution of the solution of the problem obtained by the method of characteristics described in Section 1.3. It can be confirmed that the waveform steepens in time, and a point at which the slope diverges to infinity appears at the predicted time, and the solution becomes multivalued thereafter.

1.5   SHOCK FITTING

The following are two possible ways to avoid the multivaluedness which is physically unacceptable.

(i) The occurrence of unacceptable multivaluedness means that there is some defect in the wave equation (1.16). Therefore, we return to the original physical system again and review the derivation process of (1.16), and make the necessary improvements.

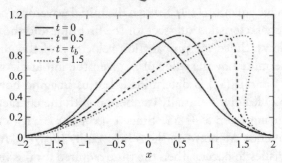

Figure 1.12: Example 3: Occurrence of multivaluedness.

For example, in the case of the traffic flow treated above, it has been assumed that the speed v of a car depends only on the density ρ at the point of space-time where the car is. However, when actually driving a car, if you know that the road ahead is congested, you may run slower than the speed corresponding to the value of ρ around you. This suggests that the velocity v can be affected not only by the density ρ there but also by its derivative ρ_x, but this point is not taken into account in the current model. (Modification of the model along this direction will be discussed in the next chapter.)

(ii) The wave equation (1.16) is derived from the conservation law and derived through a procedure that seems reasonable, so at least the part where the solution is not multivalued should have its justification. It is wasteful to abandon the whole solution just because it is multivalued in part. Then we adopt the part of the solution where it is not multivalued as it is and replace the multivalued part with an appropriate "discontinuity." Thus, we modify the original "continuous but multivalued solution" to a "single-valued but non-continuous solution."[8]

The method (i) is a serious and honest method, but the equation itself and the analysis method must become more complicated. On the other hand, although the method (ii) is rather symptomatic, it is likely to be much easier than (i). We will introduce here a method called **shock fitting** of method (ii).

When a multi-valued part as shown in Fig. 1.13 occurs in the waveform at a certain time, we would like to introduce an appropriate discontinuity and modify it to a single-valued solution. In this case, what matters is where to put the discontinuity. This problem is solved as follows.

---

[8]We have started from a differential equation (1.16). Therefore, introducing such a function with non-differentiable point will extend the class of solutions. As we shall see below, this extension is carried out based on the "conservation law in integral form," and a new solution that has a non-differentiable point introduced is called a "weak solution."

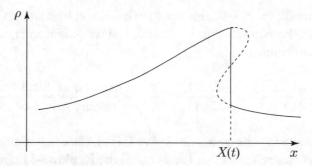

Figure 1.13: Shock fitting. Dotted line: original continuous but multi-valued solution; solid line: modified single-valued but discontinuous solution.

Although the wave equation (1.16) is defective in the sense that it gives a multi-valued solution, the conservation law of integral form

$$\frac{d}{dt} \int_a^b \rho(x,t)\, dx = q(a,t) - q(b,t),\tag{1.30}$$

which is the starting point of derivation of (1.16) is a law that must hold, and it cannot be discarded. However, since we are trying to introduce discontinuities to the solution, we cannot transform (1.30) to the conservation law of differential form (1.10) by assuming differentiability of ρ and q as before.

Let $x = X(t)$ be the location where discontinuity should be inserted at time t, $\rho(x,t)$ and $q(x,t)$ be piecewise smooth functions[9] including the discontinuity at $x = X(t)$. For an interval $[a, b]$ that contains discontinuity,

$$\int_a^b \rho(x,t)\, dx = \int_a^{X(t)} \rho(x,t)\, dx + \int_{X(t)}^b \rho(x,t)\, dx.\tag{1.31}$$

Here, in the integration on the right side, not only the integrand but also the integration interval depend on time, and the **Leibniz rule** is required to evaluate its time derivative. The Leibniz rule says that for a function $F(t)$ of t defined as an integral whose upper and lower limits also depend on t like $F(t) = \int_{a(t)}^{b(t)} f(x,t)\, dx$, the time derivative of $F(t)$ is given by

$$\frac{dF(t)}{dt} = \int_{a(t)}^{b(t)} \frac{\partial f(x,t)}{\partial t}\, dx + \dot{b}(t) f[b(t), t] - \dot{a}(t) f[a(t), t].\tag{1.32}$$

Evaluating the conservation law of integral form (1.30) using this,

$$\int_a^{X(t)} \frac{\partial \rho}{\partial t}\, dx + \dot{X}(t)\rho(X_-,t) + \int_{X(t)}^b \frac{\partial \rho}{\partial t}\, dx - \dot{X}(t)\rho(X_+,t) = q(a,t) - q(b,t),\tag{1.33}$$

[9] $f(x)$ is called "piecewise smooth" when $f(x)$ and $f'(x)$ are both piecewise continuous.

where $X_+ = X + 0$ and $X_- = X - 0$. This equation holds for any a and b. Considering the limit $a \to X_-$ and $b \to X_+$, the two integrals become 0, and we obtain an equation for the moving speed $\dot{X}(t)$ of the discontinuity,

$$\dot{X}(t) = \frac{q(X_+,t) - q(X_-,t)}{\rho(X_+,t) - \rho(X_-,t)} = \frac{\text{jump of flux } q \text{ at discontinuity}}{\text{jump of density } \rho \text{ at discontinuity}}. \tag{1.34}$$

In order to be consistent with the conservation law (1.30), the discontinuity must always move at a speed that satisfies this relationship. This is called the **Rankine–Hugoniot condition**.

In the above the condition for the speed of the discontinuity has been derived from the viewpoint of a stationary observer. The same result can also be derived more easily by introducing the viewpoint of an observer moving with the discontinuity. From the viewpoint of an observer moving at speed $\dot{X}(t)$ with the discontinuity, the discontinuity is at rest as shown in Fig. 1.14, and the densities at its front and back are $\rho(X_+,t)$ and $\rho(X_-,t)$, respectively. Since there is a loss of flux due to the discontinuity itself moving at the velocity $\dot{X}(t)$, the flux on the front side and the back side are given by $q(X_+,t) - \rho(X_+,t)\dot{X}(t)$ and $q(X_-,t) - \rho(X_-,t)\dot{X}(t)$, respectively. The discontinuity is a mere surface of thickness 0 and cannot store a finite amount of physical quantities, so at each time, the flux flowing in from the rear side must flows out from the front. Therefore, the flux on the front side and that on the back side should always be equal, that is:

$$q(X_+,t) - \rho(X_+,t)\dot{X}(t) = q(X_-,t) - \rho(X_-,t)\dot{X}(t). \tag{1.35}$$

This is nothing but the Rankine–Hugoniot condition (1.34).

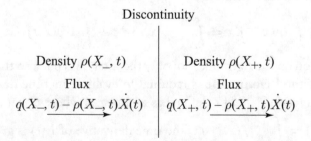

Figure 1.14: Density and flux for an observer moving with the discontinuity.

EXAMPLE 4: SPEED OF ELONGATION OF TRAFFIC JAM

On the basis of the result of the analysis performed in Example 2, find the speed of elongation of the queue of cars completely stopped in the congestion when the flow of cars of speed $v = 30$ km/h catches up with the traffic jam from behind.

[Answer]

Suppose that the tail end of the current traffic jam is $x = 0$, and there is a complete traffic jam with $\rho = \rho_{\mathrm{jam}}$ in $x > 0$, and there is a car flow of speed v in $x < 0$. From the relation between v and ρ obtained in Example 2, if ρ and q are expressed as functions of v,

$$\rho = \exp\left(\frac{v-b}{a}\right), \quad q = \rho v = v\exp\left(\frac{v-b}{a}\right), \quad a \approx -27.3\,[\mathrm{km/h}], \quad b \approx 135.7\,[\mathrm{km/h}],$$

(1.36)

and

$$\rho_{\mathrm{jam}} = \exp\left(-\frac{b}{a}\right), \quad q(\rho_{\mathrm{jam}}) = 0.$$

(1.37)

Substituting these into Rankine–Hugoniot condition (1.34) and manipulating them, the speed $\dot{X}$ of the tail of the traffic jam, which is a discontinuity point of density and flux, is given by

$$\dot{X}(t) = \frac{q(\rho_{\mathrm{jam}}) - q(v)}{\rho_{\mathrm{jam}} - \rho(v)} = \frac{-v\exp\left(\frac{v-b}{a}\right)}{\exp\left(-\frac{b}{a}\right) - \exp\left(\frac{v-b}{a}\right)}.$$

(1.38)

From this, it is estimated that when $v = 30$ km/h, the traffic jam elongates at a rate of about 15 km per hour.   ♣

....

So far, we have considered the situation in which the equation governing the wave is given by a single partial differential equation like (1.16). However, there are many types of wave phenomena that simultaneously transmit changes in multiple physical quantities, and in such a case, the governing equation is given in the form of a system of equations describing the rate of change in time of each physical quantity.

For example, when a water wave having a much longer wavelength than the water depth h, such as tsunami, propagates, the surface displacement $\eta(x,t)$ and the flow velocity $u(x,t)$ occur simultaneously. It is known that the governing equations for such a long water wave are given by[10]

$$\frac{\partial \eta}{\partial t} + \frac{\partial[(h+\eta)u]}{\partial x} = 0, \quad \frac{\partial u}{\partial t} + u\frac{\partial u}{\partial x} + g\frac{\partial \eta}{\partial x} = 0.$$

(1.39)

The method of analysis for such types of waves that simultaneously convey changes in multiple physical quantities is summarized in the Appendix B. There, it is also shown that even for this type of wave, it can be reduced to a single wave equation such as (1.16) under the condition called **the simple wave**.

For more detailed information on the overall contents of this chapter, refer to [2, 3], for example.

---

[10]For the derivation of this equation, see Appendix E.

1.6   REFERENCES

[1] H. Greenberg, An analysis of traffic flow, *Operations Research*, 7(1):79–85, 1959. 11

[2] A. J. Roberts. *A One-Dimensional Introduction to Continuum Mechanics*. World Scientific, 1994. DOI: 10.1142/2496 19

[3] G. B. Whitham. *Linear and Nonlinear Waves*. John Wiley & Sons, 1974. DOI: 10.1002/9781118032954 19

CHAPTER 2

Burgers Equation: Effect of Diffusion

If there is a temperature difference between places in an object, heat flows from high temperature to low temperature to reduce the temperature difference. When a drop of cream is dropped in a cup of coffee, the cream naturally spreads to reduce the difference in the concentration of the cream. There is an action called "diffusion" everywhere around us. In this chapter, we will consider the effect of this diffusion on wave propagation.

2.1   BURGERS EQUATION

Recall the traffic flow that was taken as an example in Chapter 1. The starting point of the story was the conservation law of cars

$$\rho_t + q_x = 0, \quad q(\rho) = \rho v(\rho), \tag{2.1}$$

where subscripts t and x denote partial derivatives with respect to them. The car speed v is a monotonically decreasing function of the car density ρ, and the simplest approximation to $v(\rho)$ would be to assume a linear function of ρ. Then,

$$v = v_0(1 - \rho/\rho_{\text{jam}}) \quad \longrightarrow \quad q = v_0(\rho - \rho^2/\rho_{\text{jam}}), \tag{2.2}$$

and (2.1) reduces to a nonlinear wave equation like

$$\rho_t + v_0\rho_x + \alpha\rho\rho_x = 0, \quad \alpha = -2v_0/\rho_{\text{jam}}. \tag{2.3}$$

As we studied in Chapter 1, the solution of (2.3) becomes multivalued because of the intersection of characteristics, and if we do not introduce an artificial "discontinuity," the solution breaks down at some finite time.

The traffic model assumes that the speed v of the car is determined by the local density ρ at that time. However, when actually driving a car, the driver always pays attention to the road conditions ahead as well as around him. For example, if you think that the road ahead is crowded and you will not be able to run smoothly there, you may run at a somewhat modest speed even if your surroundings allow you to run faster, and conversely if you know that the road ahead is free and the flow is faster, you may run faster than the speed determined from your surroundings, expecting the next acceleration.

One way to introduce such a "look ahead" effect in the traffic flow model is to add a term proportional to the spatial derivative ρ_x of ρ to the equations of v and q, such as,

$$v = v_0(1 - \rho/\rho_{\text{jam}}) - \frac{v}{\rho}\rho_x \quad \longrightarrow \quad q = v_0(\rho - \rho^2/\rho_{\text{jam}}) - v\rho_x. \tag{2.4}$$

Here the coefficient v of the additional term must be positive in order to reduce the speed when the road ahead is more crowded ($\rho_x > 0$). Then the conservation law of the car (2.1) becomes

$$\rho_t + v_0\rho_x + \alpha\rho\rho_x = v\rho_{xx}, \qquad \alpha = -2v_0/\rho_{\text{jam}}, \tag{2.5}$$

and a second derivative term ρ_{xx} is newly added.

In the context of fluid mechanics, Lighthill [4] showed, by a systematic analysis using a perturbation method starting from a system of basic equations for gases, and by taking into consideration of the dissipative effects such as viscosity and heat conduction, that the propagation of weakly nonlinear sound waves is governed by

$$v_t + c_0v_x + vv_x = \delta v_{xx}, \tag{2.6}$$

which is an equation with exactly the same form as (2.5).[1]

In fluid mechanics, the nonlinear PDE

$$u_t + uu_x = vu_{xx} \tag{2.7}$$

is well known as the **Burgers equation**. This equation was proposed in 1939 by Burgers as a simple 1D model of the Navier–Stokes equation

$$u_t + (u \cdot \nabla)u = -\frac{1}{\rho}\nabla p + v\nabla^2 u, \tag{2.8}$$

which is the most basic equation of motion of fluid mechanics, in order to study turbulence phenomena [1]. Here, $u(x, t)$ is the flow velocity vector, $p(x, t)$ is the pressure, and v is a physical constant called the "kinematic viscosity" of the fluid. In this book, not only (2.7) but also all nonlinear PDF of the form

$$v_t + c_0v_x + \alpha vv_x = vv_{xx}, \tag{2.9}$$

like (2.5) and (2.6), are called the "Burgers equation." The term c_0v_x in (2.9) can always be eliminated by looking at the system from a framework translating at speed c_0. That is, if new space variable ξ and time variable τ are introduced by

$$\xi = x - c_0t, \quad \tau = t, \tag{2.10}$$

---

[1]Here, c_0 represents the sound velocity in the stationary gas, and $v(x, t)$ represents $v = u + c - c_0$, that is, the deviation of the propagation velocity $u + c$ of the C_+ characteristics from the static sound velocity c_0. (For more detail, see Appendix B.) δ is a physical constant called "acoustic diffusivity" which is determined from the viscosity coefficient and the thermal diffusivity of the gas, and in the case of air at normal temperature, it takes a very small value of about 2×10^{-5} m^2/s.

and $v(x, t)$ is treated as a function of ξ and τ, according to the chain rule of partial differentiation

$$\frac{\partial}{\partial t} = \frac{\partial}{\partial \tau} \frac{\partial \tau}{\partial t} + \frac{\partial}{\partial \xi} \frac{\partial \xi}{\partial t} = \frac{\partial}{\partial \tau} - c_0 \frac{\partial}{\partial \xi}, \tag{2.11a}$$

$$\frac{\partial}{\partial x} = \frac{\partial}{\partial \tau} \frac{\partial \tau}{\partial x} + \frac{\partial}{\partial \xi} \frac{\partial \xi}{\partial x} = \frac{\partial}{\partial \xi}, \tag{2.11b}$$

(2.9) can be rewritten as

$$v_\tau + \alpha v v_\xi = \nu v_{\xi\xi}. \tag{2.12}$$

At the same time, if we introduce a new dependent variable $u = \alpha v$, (2.9) gives

$$u_\tau + u u_\xi = \nu u_{\xi\xi}, \tag{2.13}$$

i.e., the original form of the Burgers equation (2.7). From this, all PDE of the form (2.9) is equivalent to the Burgers equation (2.7). From now on, we will consider only the Burgers equation of the form[2]

$$u_t + u u_x = \nu u_{xx}. \tag{2.14}$$

2.2   DIFFUSION EFFECT

The second derivative term νu_{xx} of the Burgers equation is called the **diffusion term**. In nature, when there is a spatial non-uniformity in the density of a physical quantity, it is often observed that the flux of the quantity is automatically generated so as to eliminate the non-uniformity, and is generally called the **diffusion phenomenon**. For example, heat flows from a high temperature part to a low temperature part, and a drop of a cream dropped in a cup of coffee spreads around. These are all diffusion phenomena.

Taking the temperature distribution of a straight wire as an example, let us consider the equation governing the diffusion phenomenon more specifically.[3] Take the x axis along the wire, and let $u(x, t)$ [K] be the temperature at point x and time t. Assuming that the heat capacity per 1 m of the wire is α [J/m K], the density of thermal energy $\rho(x, t)$ [J/m] can be written as $\rho = \alpha u$. If the flux of thermal energy is denoted by $q(x, t)$ [J/s], the conservation law of thermal energy requires the equation

$$\rho_t + q_x = 0 \tag{2.15}$$

to hold. The heat flux $q(x, t)$ flows from the high temperature to the low temperature, and its magnitude is larger as the temperature changes more rapidly. The simplest mathematical model to express this property is to assume

$$q = -\beta u_x, \tag{2.16}$$

---

[2]If we change the space and time scale by $\tilde{t} = t/\nu$, $\tilde{x} = x/\nu$, we can transform (2.14) to $u_{\tilde{t}} + u u_{\tilde{x}} = u_{\tilde{x}\tilde{x}}$ and can treat the diffusion coefficient ν as 1, but we will not do it here.

[3]In Fig. 1.2 of Chapter 1, it is sufficient to replace the electric charge with the unit of coulomb with the thermal energy with the unit of Joule.

which is called "Fourier's law" of heat conduction. Here, β is a positive material constant called the "heat conduction coefficient." Substituting $\rho = \alpha u$ and (2.16) into (2.15) immediately gives the **heat equation**

$$u_t = \nu\, u_{xx}, \qquad (\nu = \beta/\alpha), \tag{2.17}$$

which governs the evolution of the temperature distribution $u(x, t)$.

In the above, (2.17) was derived in the context of temperature change due to heat conduction. But for its derivation, we have used only very natural laws such as conservation law (2.15) and Fourier's law (2.16). Reflecting this fact, (2.17) holds true as it is if $u(x, t)$ is not the temperature but some other quantity such as the concentration of solute in a certain solution. Thus, (2.17) is one of the most important PDEs that appear in various fields of natural science, and in more general context, it is called the **diffusion equation**.

Figure 2.1 shows an example of the behavior of the solution of (2.17), in which we can see how $u(x, t)$, which was initially concentrated as a sharp pulse, becomes more and more flat with time.

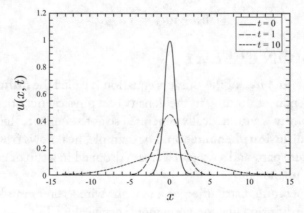

Figure 2.1: Behavior of a solution of the diffusion equation ($\nu = 1$, $u(x, 0) = e^{-x^2}$).

As we studied in Chapter 1, the term $u u_x$ on the left side of the Burgers equation (2.14) represents the nonlinear effect that the part of the waveform with larger u propagates faster, so that the waveform will lean forward and steepens. On the other hand, the diffusion term νu_{xx} on the right side tends to eliminate the spatial difference in u as discussed above. Thus, the Burgers equation is a concise and interesting wave equation that has two competing effects, the nonlinear effect that makes the waveform steeper and the diffusion effect that makes it flatter. In the Burgers equation, since the diffusion term always stops the steepening by the nonlinear term, multivalued solutions will not appear for any initial waveform.

The last term $\nu \nabla^2 \boldsymbol{u}$ of the Navier–Stokes equation (2.8) is usually called the "viscosity term," but it physically corresponds to the 3D version of the diffusion term νu_{xx} discussed above. If there is non-uniformity in momentum (i.e., spatial difference in flow velocity $\boldsymbol{v}$), momentum flux is automatically generated from the faster part to the slower part to eliminate the velocity difference, and as a result of this the slower part is accelerated and the faster part is decelerated. We usually call this diffusion of momentum "viscosity."

It is known that the water surface displacement $\eta(x, t)$ is governed by the equation

$$\eta_t + \left(3\sqrt{g(h + \eta)} - 2\sqrt{gh}\right)\eta_x = 0, \tag{2.18}$$

when the wavelength of the water wave is much longer than the water depth h. (For more detail, see Appendices B and F.) As the waves rush to the coast, they gradually steepen and start to break up. This is called **wave breaking**. Wave breaking causes energy dissipation, but this effect is not taken into consideration in (2.18). In (2.18), if $\sqrt{g(h + \eta)}$ is approximated by $\sqrt{gh}(1 + \eta/2h)$ by assuming that η is small compared to the water depth h, and if the dissipation of wave energy by wave breaking is modeled by the diffusion term η_{xx} by analogy from the viscosity term of the Navier–Stokes equation, (2.18) becomes

$$\eta_t + c_0\eta_x + \alpha\eta\eta_x = \nu\eta_{xx}, \quad c_0 = \sqrt{gh}, \quad \alpha = \frac{3}{2h}\sqrt{gh}, \tag{2.19}$$

and the Burgers equation appears again. In the field of coastal engineering, this equation is sometimes used as a simple model equation for waves in the wave breaking zone near the coast.

2.3 HOPF–COLE TRANSFORMATION: CLOSE RELATION TO DIFFUSION EQUATION

The Burgers equation (2.14) is an unusual nonlinear PDE in the sense that its initial value problem can be solved analytically. Equation (2.14) can be transformed into

$$(-u)_t = \left(\frac{1}{2}u^2 - \nu u_x\right)_x. \tag{2.20}$$

This means that there exists a scalar function $\phi(x, t)$ such that

$$\phi_x = -u, \qquad \phi_t = \frac{1}{2}u^2 - \nu u_x. \tag{2.21}$$

Then, deleting u from (2.21) gives

$$\phi_t = \frac{1}{2}(\phi_x)^2 + \nu\phi_{xx}. \tag{2.22}$$

If we introduce ψ such that $\phi = 2\nu \ln \psi$ (ln is a natural logarithm), the above equation is transformed into

$$2\nu\frac{\psi_t}{\psi} = \frac{1}{2}\left(2\nu\frac{\psi_x}{\psi}\right)^2 + \nu\left(2\nu\frac{\psi_{xx}\psi - \psi_x^2}{\psi^2}\right) \quad \longrightarrow \quad \psi_t = \nu\psi_{xx}. \tag{2.23}$$

That is, the Burgers equation (2.14) for u can be converted into a more manageable linear PDE, the diffusion equation, for ψ, by using the variable transformation

$$u = -2v\frac{\psi_x}{\psi}. \tag{2.24}$$

The transformation used here is called **Hopf–Cole transformation** from the name of the researchers who first found it [2, 3].

It is known that the analytical solution of the initial value problem of the diffusion equation (2.23) in the infinite domain is given by

$$\psi(x,t) = \frac{1}{\sqrt{4\pi vt}} \int_{-\infty}^{\infty} \psi_0(x') \exp\left[-\frac{(x-x')^2}{4vt}\right] dx', \tag{2.25}$$

where $\psi_0(x)$ represents the distribution of $\psi(x,t)$ at $t = 0$.[4] Therefore, the initial value problem of the Burgers equation can be solved by the following procedure.

1. Find $\psi_0(x)$ corresponding to the given initial condition $u(x,0)$ from (2.24).

2. Find $\psi(x,t)$ at an arbitrary time t from (2.25).

3. Find $u(x,t)$ corresponding to this $\psi(x,t)$ from (2.24).

2.4　TYPICAL SOLUTIONS OF THE BURGERS EQUATION

The Burgers equation (2.14) is a nonlinear equation, and even if $u_1(x,t)$ and $u_2(x,t)$ are both functions satisfying (2.14), their linear combination does not satisfy (2.14). On the other hand, the diffusion equation (2.23) is a homogeneous linear differential equation, and if $\psi_1(x,t)$ and $\psi_2(x,t)$ satisfy (2.23), then any linear combination $c_1\psi_1(x,t) + c_2\psi_2(x,t)$ also satisfies (2.23). Using this linearity and the correspondence between the diffusion equation and the Burgers equations shown above, we can easily find some basic solutions of the Burgers equation as follows:

2.4.1　UNIFORM SOLUTION

For any constant u_1, c_1,

$$\psi(x,t) = \exp\left[-\frac{u_1}{2v}x + \frac{u_1^2}{4v}t - c_1\right] \tag{2.26}$$

is obviously a solution of the diffusion equation (2.23). Therefore, $u(x,t)$ connected with this by Hopf–Cole transformation (2.24), i.e.,

$$u = -2v\frac{\psi_x}{\psi} = -2v\frac{-\frac{u_1}{2v}\psi}{\psi} = u_1 \, (= \text{constant}) \tag{2.27}$$

[4]For the derivation of this solution, see Appendix C.

is a solution of the Burgers equation. The fact that a constant is a solution of the Burgers equation can be understood from a look at the equation itself, and this is a trivial and boring solution. However, this trivial solution turns out to be surprisingly useful in finding more complex solutions as shown below.

2.4.2 SHOCK WAVE SOLUTION

Because the diffusion equation (2.23) is homogeneous and linear, the principle of superposition of solutions holds. That is, the sum $\psi = \psi_1 + \psi_2$ of two solutions

$$\psi_1 = \exp\left[-\frac{u_1}{2\nu}x + \frac{u_1^2}{4\nu}t - c_1\right], \quad \psi_2 = \exp\left[-\frac{u_2}{2\nu}x + \frac{u_2^2}{4\nu}t - c_2\right], \tag{2.28}$$

also a solution of (2.23). The solution $u(x,t)$ of the Burgers equation connected to this ψ by the Hopf–Cole transformation is given by

$$u = -2\nu\frac{\psi_x}{\psi} = -2\nu\frac{-\frac{u_1}{2\nu}\psi_1 - \frac{u_2}{2\nu}\psi_2}{\psi_1 + \psi_2} = \frac{u_1\psi_1 + u_2\psi_2}{\psi_1 + \psi_2}. \tag{2.29}$$

If we introduce θ and x_0 so that

$$\begin{aligned}
\frac{\psi_1}{\psi_2} &= \exp\left[-\frac{1}{2\nu}(u_1 - u_2)x + \frac{1}{4\nu}(u_1^2 - u_2^2)t - (c_1 - c_2)\right] \\
&= \exp\left[\frac{1}{2\nu}(u_2 - u_1)\left(x - \frac{u_1 + u_2}{2}t - x_0\right)\right] = \exp\theta,
\end{aligned} \tag{2.30}$$

then $u(x,t)$ can be rewritten as

$$\begin{aligned}
u &= \frac{u_1 e^{\theta/2} + u_2 e^{-\theta/2}}{e^{\theta/2} + e^{-\theta/2}} = \frac{1}{2}(u_1 + u_2) - \frac{1}{2}(u_2 - u_1)\frac{e^{\theta/2} - e^{-\theta/2}}{e^{\theta/2} + e^{-\theta/2}} \\
&= \frac{1}{2}(u_1 + u_2) - \frac{1}{2}(u_2 - u_1)\tanh\frac{\theta}{2} \\
&= \frac{1}{2}(u_1 + u_2) - \frac{1}{2}(u_2 - u_1)\tanh\left[\frac{u_2 - u_1}{4\nu}\left(x - \frac{u_1 + u_2}{2}t - x_0\right)\right],
\end{aligned} \tag{2.31}$$

where $\tanh x$ is a hyperbolic function defined by $\tanh x = (e^x - e^{-x})/(e^x + e^{-x})$, which is a monotonically increasing odd function connecting the value -1 at $x \to -\infty$ and the value $+1$ at $x \to \infty$. If $u_1 < u_2$, (2.31) has a waveform that smoothly connects the constant state $u = u_2$ at $x \to -\infty$ and the constant state u_1 at $x \to \infty$ as shown in Fig. 2.2, and translates at a constant speed $c = (u_1 + u_2)/2$ without changing the waveform.[5] This is called the **shock wave solution** of the Burgers equation.

[5]Regardless of the magnitude relationship between u_1 and u_2, (2.31) asymptotes to $u = \max(u_1, u_2)$ as $x \to -\infty$ and to $u = \min(u_1, u_2)$.

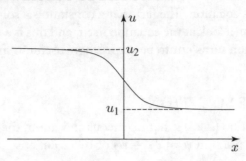

Figure 2.2: Shock wave solution of the Burgers equation.

The representative value d of the width of the shock wave, that is, the distance of transition from u_2 to u_1 is given by

$$d = \frac{4\nu}{|u_2 - u_1|}. \tag{2.32}$$

Therefore, the stronger the shock wave (that is, the greater the jump $|u_2 - u_1|$) and the smaller the diffusion coefficient ν, the thinner the shock wave, and in the limit $\nu \to 0$ the shock wave becomes a discontinuity of thickness zero.

The fact that the thickness of the shock wave solution is determined as (2.32) indicates that this shock wave solution is realized on the balance between the nonlinear effect and the diffusion effect. If there is only a nonlinear effect, the waveform will lean forward and steepens, and the solution will break down. Also, if there is only a diffusion effect, any waveform will become flatter with time, and steady state will not be realized. In order for the nonlinear effect that steepens the waveform to be balanced with the diffusion effect that flattens the waveform, the magnitudes of both must be comparable. If the thickness of the shock wave is d and the jump of u between front and rear of the shock wave is U, the magnitudes of the nonlinear term uu_x and the diffusion term νu_{xx} are estimated as $O(U^2/d)$ and $O(\nu U/d^2)$, respectively. It can be seen that the d determined by (2.32) is exactly the thickness to make the magnitudes of these two competing terms comparable.

In the limit $\nu \to 0$, the Burgers equation becomes a single hyperbolic equation

$$u_t + uu_x = 0, \text{that is} u_t + q_x = 0, \quad q := \frac{1}{2}u^2. \tag{2.33}$$

In this hyperbolic equation, the characteristics catches up when there is a portion of $u_x < 0$ in the initial condition, and the solution becomes multivalued. In the method of "shock fitting" discussed in Chapter 1, we introduced an artificial discontinuity to the solution to eliminate this multivaluedness. In the case of (2.33), according to the Rankine–Hugoniot condition (1.34), the discontinuity between the states with $u = u_1$ and $u = u_2$ should propagate at the speed $\dot{X}$

given by

$$\dot{X} = \frac{\frac{1}{2}u_2^2 - \frac{1}{2}u_1^2}{u_2 - u_1} = \frac{u_1 + u_2}{2}, \tag{2.34}$$

which matches the propagation velocity of the shock wave solution of the Burgers equation found above. From this, it can be said that the shock wave solution of the Burgers equation asymptotes to the discontinuity which is artificially introduced to the hyperbolic equation in the $\nu \to 0$ limit, and when $\nu \neq 0$, it expresses the internal structure (i.e., the state of transition) of the shock wave with finite thickness.

EXAMPLE 1: STEADILY TRAVELING SOLUTION OF THE BURGERS EQUATION

Find the shock wave solution (2.31) as the steady traveling wave solution of the Burgers equation (2.14).

[Answer]

A steady traveling wave solution is a solution that travels at a constant speed without changing the waveform. For the linear wave equation $u_t + c_0 u_x = 0$ (c_0 =constant), whatever waveform $F(x)$ is given initially, it only translates at velocity c_0, so for an arbitrary function $F(x)$, $u(x, t) = F(x - c_0 t)$ is a steady traveling solution. However, in the case of the Burgers equation, which is nonlinear, the waveform given initially hardly translates as it is except for a special waveform.

In general, when a PDE for $u(x, t)$ is given, its steady traveling solution can be obtained by the following procedure.

1. Assuming that the initial waveform of the steady traveling wave solution is $U(x)$ and the propagation velocity is c, we can write $u(x, t) = U(x - ct)$. Of course neither $U(x)$ nor c are known yet.

2. Introduce ξ by $\xi = x - ct$. When $u(x, t)$ can be written as $U(x - ct)$, u depends on x and t only through ξ. Therefore,

$$\frac{\partial u}{\partial t} = \frac{dU}{d\xi} \frac{\partial \xi}{\partial t} = -c \frac{dU}{d\xi}, \quad \frac{\partial u}{\partial x} = \frac{dU}{d\xi} \frac{\partial \xi}{\partial x} = \frac{dU}{d\xi}. \tag{2.35}$$

Substituting these into the original PDE for $u(x, t)$ gives an ordinary differential equation for $U(\xi)$.

3. Find $U(\xi)$ by solving this ordinary differential equation with appropriate boundary conditions.

Let us apply this procedure to the Burgers equation to find the shock wave solution (2.31). Assuming $u(x, t) = U(x - ct)$ and substituting into the Burgers equation (2.14), we obtain an ordinary differential equation for $U(\xi)$ as follows:

$$-cU' + UU' = \nu U'',\tag{2.36}$$

where U' stands for $dU/d\xi$. Integrating both sides once with respect to ξ,

$$\frac{1}{2}U^2 - cU = \nu U' + c_1,\tag{2.37}$$

where c_1 is the integration constant. Assuming a shock wave solution, we employ the boundary conditions $U \to u_2$ $(\xi \to -\infty)$, $U \to u_1$ $(\xi \to +\infty)$. Since U approaches a constant value at $\xi \to \pm\infty$, $U' \to 0$ $(\xi \to \pm\infty)$. Evaluating (2.37) at $\xi \to \pm\infty$,

$$\frac{1}{2}u_2^2 - cu_2 = c_1, \quad \frac{1}{2}u_1^2 - cu_1 = c_1,\tag{2.38}$$

which immediately gives

$$c = \frac{1}{2}(u_1 + u_2), \quad c_1 = -\frac{1}{2}u_1 u_2.\tag{2.39}$$

The property of the shock wave solution (2.31) that the propagation velocity c is the average of u_1 and u_2 is thus obtained. Substituting these c and c_1 into (2.37) yields,

$$\frac{dU}{d\xi} = \frac{1}{2\nu}(U - u_1)(U - u_2).\tag{2.40}$$

This is a separable first order differential equation and can be solved as follows:

$$\frac{dU}{(U - u_1)(U - u_2)} = \frac{d\xi}{2\nu} \quad \longrightarrow \quad \left(\frac{1}{U - u_2} - \frac{1}{U - u_1}\right)dU = \frac{u_2 - u_1}{2\nu}d\xi$$

$$\longrightarrow \quad \ln\left|\frac{U - u_2}{U - u_1}\right| = \frac{u_2 - u_1}{2\nu}(\xi - x_0),\tag{2.41}$$

where x_0 is the integration constant. If we denote the right side of (2.41) as θ, and remove the absolute value by taking care that $u_1 < U < u_2$, we obtain

$$\frac{u_2 - U}{U - u_1} = e^\theta \quad \longrightarrow \quad U = \frac{u_1 e^{\theta/2} + u_2 e^{-\theta/2}}{e^{\theta/2} + e^{-\theta/2}},\tag{2.42}$$

and the rest will be the same as (2.31).   ♣

. . . .

2.4.3   COALESCENCE OF SHOCK WAVES

The sum of the three basic solutions of the diffusion equation $\psi = \psi_1 + \psi_2 + \psi_3$ with

$$\psi_i = \exp\left[-\frac{u_i}{2\nu}x + \frac{u_i^2}{4\nu}t - c_i\right], \quad (i = 1, 2, 3)\tag{2.43}$$

is also a solution of the diffusion equation (2.23), and the solution of the Burgers equation connected to this by the Hop–Cole transformation is given by

$$u = -2v\frac{\psi_x}{\psi} = \frac{u_1\psi_1 + u_2\psi_2 + u_3\psi_3}{\psi_1 + \psi_2 + \psi_3}. \tag{2.44}$$

The concrete expression of this solution $u(x, t)$ becomes somewhat complicated, so it is not described here. But, assuming $u_1 < u_2 < u_3$, and if the relationship of the initial positions are appropriate, this solution describes the process of shock wave coalescence in which the faster shock wave (velocity$= \frac{u_2+u_3}{2}$) connecting u_3 and u_2 in the rear catches up the slower shock wave (velocity$= \frac{u_1+u_2}{2}$) connecting u_2 and u_1 in front, and finally turns into one stronger shock wave directly connecting u_3 and u_1 (see Fig. 2.3). As Lighthill [4] showed, the Burgers equation is an equation that describes the propagation of (weak) shock waves in the real air, so this solution tells us that the shock waves that travel through the air actually have the property that they coalesce when they catch up and form a stronger shock wave.

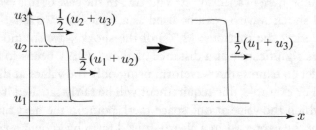

Figure 2.3: Coalescence of two shocks.

EXAMPLE 2: NUMERICAL SIMULATION OF THE BURGERS EQUATION

Create a program that numerically traces the evolution of the Burgers equation. Also, using it, reproduce by numerical simulation the process of coalescence in which the shock wave S_1 connecting $u = 3$ and $u = 2$ catches up with the shock wave S_2 connecting $u = 2$ and $u = 0$ to form one large shock wave S_3 connecting $u = 3$ and $u = 0$. Let the value of the diffusion coefficient v be 0.1.

[Answer]

The Burgers equation (2.14) can be written as

$$\frac{\partial u}{\partial t} + \frac{\partial F}{\partial x} = v\frac{\partial^2 u}{\partial x^2}, \quad F := \frac{1}{2}u^2. \tag{2.45}$$

As an example of the simplest finite difference approximations of (2.45), there is the following numerical scheme that approximates the time derivative with forward difference and the space derivative with central difference:

$$\frac{u_j^{n+1} - u_j^n}{\Delta t} + \frac{F_{j+1}^n - F_{j-1}^n}{2\Delta x} = v\frac{u_{j-1}^n - 2u_j^n + u_{j+1}^n}{(\Delta x)^2}. \tag{2.46}$$

Here, Δt and Δx are the intervals for discretization of time and space, respectively, and u_j^n denotes the value of u at nth time step and jth space mesh point. The unknown in (2.46) is only u_j^{n+1}, and the following can be obtained:

$$u_j^{n+1} = u_j^n - \frac{\Delta t}{4\Delta x}\left[\left(u_{j+1}^n\right)^2 - \left(u_{j-1}^n\right)^2\right] + \frac{v\Delta t}{\Delta x^2}\left(u_{j-1}^n - 2u_j^n + u_{j+1}^n\right). \tag{2.47}$$

By using (2.47), if the value of u at all space mesh points are known at the nth time step, then the approximate value of u at each space mesh point in the $(n + 1)$th time step can be obtained, and by repeating this process, the evolution of the waveform can be traced once it is given at $t = 0$.

Care must be taken here in selecting Δt and Δx. In the case of this example, the thickness of d of the coalesced shock wave S_3 can be used as a representative length scale of the waveform change. Considering the thickness (2.32) of the shock wave solution, $d = 0.4/3$. Since the waveform changes significantly at a distance of about d, Δx needs to be sufficiently short compared to d in order to express the waveform using only the values at discrete points with a reasonable accuracy. For example, this requirement will be fairly satisfied if $\Delta x = d/10$, say.

A scheme in which the value at one space mesh point in the next time step, which is an unknown quantity, can be expressed in a clearly solved form by known quantities, as in (2.47), is called an **explicit scheme**. In an explicit scheme, numerical calculations may "explode" and do not work unless the time step Δt is chosen small enough in conjunction with the space step Δx. In the case of this example, when Δx is determined to a certain value as described above, the numerical calculation will not work unless we specify Δt so that both the coefficients of the second term $\frac{\Delta t}{4\Delta x}$ and the third term $\frac{v\Delta t}{\Delta x^2}$ are sufficiently small like 0.1, say.[6]

Figure 2.4 is an example of the numerical result obtained by the scheme (2.47). At $t = 0$, the centers of S_1 and S_2 are located at $x = 5$ and $x = 15$, respectively. Since the distance between the two is far enough initially to ignore each other's interference, they initially travel as single shock waves. According to the shock wave solution (2.31), the speeds of S_1 and S_2 are 2.5 and 1, respectively, so the two are expected to coalesce around $x \approx 21.7$ at $t \approx 6.7$ and become a single large shock wave S_3 that directly connects 3 and 0 and propagates at a speed of 1.5. It can be seen that the results of the numerical calculation shown in Fig. 2.4 correctly reproduces all such theoretical predictions.

♣

. . . .

[6]This is a requirement from the "stability" of the numerical scheme. For more details of this topic, refer to textbooks on numerical analysis.

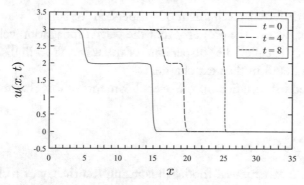

Figure 2.4: Numerical results of shock wave coalescence.

2.4.4   BORE

Similar to the shock waves in gases, there is a water wave phenomenon in which a waveform accompanied by a sudden change in water surface height travels for a long distance, and is called a **bore**. Bores can be seen worldwide in rivers with large difference in tides at the estuary. The Pororoca of the Amazon, the bore of the River Severn in England, and that of Fundy Bay of Canada are among the most famous ones.

Figure 2.5: Bore in the River Severn (photographed by the author).

When energy dissipation is not taken into consideration, the long wave propagation to the still water is governed by the nonlinear wave equation such as (2.18), and the nonlinear term brings about the steepening of the waveform. The steepening is balanced with the energy dissipation due to wave breaking, and a bore that propagates over long distance is realized. The Burgers equation (2.19) may be used as a simplest model for this phenomenon of water waves, too.

However, there is one big difference between bores in water waves and shock waves in gases. In the shock waves in gases, it is energy dissipation due to diffusion effects such as viscosity and heat conduction that stop the steepening of the waveform. On the other hand, in the case of bores of water waves, in addition to such diffusion effects (i.e., energy dissipation due to breaking), a totally different mechanism of "dispersion" of surface waves also contributes to energy consumption of bores. The property that the wave propagation velocity depends on the wavelength or frequency is called the "dispersion" of the wave. We will discuss the dispersion of water surface waves in detail in the next chapter.

For more detailed information on the overall contents of this chapter, refer to [4] and [5], for example.

2.5   REFERENCES

[1] J. M. Burgers. A mathematical model illustrating the theory of turbulence. *Advances in Applied Mechanics*, 1:171–199, 1948. DOI: 10.1016/s0065-2156(08)70100-5 22

[2] J. D. Cole. On a quasilinear parabolic equation occuring in aerodynamics. *Quarterly of Applied Mathematics*, 9:225–236, 1951. DOI: 10.1090/qam/42889 26

[3] E. Hopf. The partial differential equation $u_t + uu_x = \mu u_{xx}$. *Communications on Pure and Applied Mathematics*, pp. 201–230, 1950. DOI: 10.1002/cpa.3160030302 26

[4] M. J. Lighthill. Viscosity effects in sound waves of finite amplitude. In G. K. Batchelor and R. M. Davies, Eds., *Surveys in Mechanics*, pp. 250–351, Cambridge University Press, 1956. 22, 31, 34

[5] G. B. Whitham. *Linear and Nonlinear Waves*. John Wiley & Sons, 1974. DOI: 10.1002/9781118032954 34

CHAPTER 3

Basics of Linear Water Waves

As stated in the Preface, this book is aimed at an introduction to physical aspects of various nonlinear phenomena of waves in general. However, it is beneficial to use some concrete wave to explain various phenomena and the mechanisms behind them. So we will take the water waves as a suitable example for this purpose which is ubiquitous in our everyday life. Water wave is totally different from light and sound waves in that it propagates at a different speed depending on the wavelength and frequency. This property is called "dispersion." The dispersion plays crucial roles in all of the various wave phenomena treated in the rest of this book. In this chapter we mainly focus on the dispersion of waves, and will neglect the effect of nonlinearity by assuming that the amplitude of the wave is infinitesimally small.

3.1   DISPERSION RELATION

Let the temporal and spatial development of some physical quantity $u(x, t)$ be governed by a PDE. Then the first thing to do to study the wave phenomena in this system is to find the **linear sinusoidal wave solution** like

$$u(x, t) = A \cos(kx - \omega t + \theta_0). \tag{3.1}$$

This is the most basic element of any kind of wave phenomenon.

Suppose that there is a long wave flume equipped with a wave maker at one end, and the side of the tank is transparent so that waves can be seen from the side (see Fig. 3.1). If a photograph of the water surface waveform is taken through the side at a certain time, the waveform seen is a function of only x, and it is called **the spatial waveform**. The spatial waveform $u(x, t_0)$ at time t_0 is given by

$$u(x, t_0) = A \cos(kx + \theta_1), \quad \theta_1 = -\omega t_0 + \theta_0 \, (= \text{const.}). \tag{3.2}$$

The spatial distance between adjacent crests or troughs is the **wavelength** and is denoted here by λ [m]. The wavelength is the spatial distance required for the phase to change by 2π, so the relation $\lambda = 2\pi/k$ holds.[1] From this, k expresses the number of waves in 2π [m] and is called **wavenumber**. The unit of k is [rad/m].

---

[1]In general, the expression inside the parentheses of $\cos(\cdots)$ or $\sin(\cdots)$ is called the **phase**.

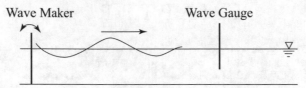

Figure 3.1: A wave flume equipped with a wave maker.

Next, suppose that a wave gauge is installed at some fixed location along the wave flume to observe the rise and fall of the water surface there as waves go through the location. The wave-form thus measured is a function of only t and is called the **temporal waveform**. The temporal waveform $u(x_0, t)$ at $x = x_0$ is given by

$$u(x_0, t) = A \cos(\omega t + \theta_2), \quad \theta_2 = -k x_0 - \theta_0 \ (= \text{const.}). \tag{3.3}$$

The temporal distance between adjacent crests or troughs in the temporal waveform is the **period** and is denoted here by T [s]. The period is the interval of time required for the phase to change by 2π, so there is a relation of $T = 2\pi/\omega$ between T and ω. Thus, ω expresses the number of waves in 2π [s] and is called **angular frequency**. The unit of ω is [rad/s].[2]

The k and ω need to satisfy some specific relation in order for (3.1) to satisfy the governing PDE (and also the boundary conditions if imposed). This relation is called the **linear dispersion relation** of the system. For example, let us imagine a train of waves which is generated by moving the wave maker at some fixed frequency. If you move the wave maker quickly, short waves would be generated, and if you move it slowly, long waves would be generated and transmitted through the wave flume. This clearly indicates that k and ω are linked with each other in order to satisfy the governing equation of the system. The specific functional form of the linear dispersion relation of course depends on the individual system. For example, the electro-magnetic wave has its own dispersion relation which is totally different from that of the surface water waves, as shown below.

The position of a crest of the spatial waveform of the linear sinusoidal wave (3.1) at $t = t_0$ is a point where the phase $kx - \omega t_0 + \theta_0$ is exactly equal to $2m\pi$ (m is integer). Let us focus on a particular crest whose phase corresponds to $2m\pi$. Even if the position of this crest moves after a while, the value of the phase at that point remains $2m\pi$ in order the crest remains to be so. Therefore, assuming that the crest moves a small distance Δx during a short time Δt,

$$kx - \omega t_0 + \theta_0 = k(x + \Delta x) - \omega(t_0 + \Delta t) + \theta_0 = 2m\pi, \tag{3.4}$$

that is

$$k\Delta x - \omega \Delta t = 0. \tag{3.5}$$

---

[2]The frequency of f [Hz] means the number of oscillation in one second, and is related to ω by $f = \omega/(2\pi)$.

From this it can be seen that the wave velocity c is given by

$$c = \frac{\Delta x}{\Delta t} = \frac{\omega}{k}. \tag{3.6}$$

This wave velocity is the velocity at which the wave phase propagates, and it is called **phase velocity** in distinction from the group velocity described later.

Rewriting (3.1) as

$$u(x,t) = A \cos\left[k\left(x - \frac{\omega}{k}t\right) + \theta_0\right], \tag{3.7}$$

it can be seen that u, which is a function of x and t, depends only on the combination $\xi = x - (\omega/k)t$. This indicates that $u(x,t)$ is translating at a speed of ω/k, as we studied at the beginning of Chapter 1, and this also explains that the wave speed is given by ω/k. Furthermore, this expression for c can also be obtained from the fact that the wave travels one wavelength during one period, that is,

$$c = \frac{\lambda}{T} = \frac{2\pi/k}{2\pi/\omega} = \frac{\omega}{k}. \tag{3.8}$$

Since the units of ω and k are [rad/s] and [rad/m], respectively, c given by ω/k has unit of [m/s] as it should.

The linear dispersion relation makes ω a function of k. Therefore, the wave velocity c given by (3.6) is also generally a function of k except in the special case that ω is directly proportional to k. According to the Fourier analysis, any initial waveform can be considered to be a superposition of sinusoidal waves of various wavelengths.[3] If the wave velocity $c(k)$ depends on k and hence on the wavelength λ, the various wavelength components that make up the initial waveform propagate at different speeds, and as a result the waveform will change with time. Even if the initial waveform is localized in a narrow space, the waveform will be scattered more and more with time. From this, a wave whose $c(k)$ is not constant but depends on k is called **dispersive wave**.

As described in the next section, the water wave is dispersive. Since the wave has the property of dispersion except in the special case in which the dispersion relation is given by $\omega = c_0 k$ (c_0 is a constant), there are many kinds of dispersive waves other than water surface waves, for example, global-scale waves called Rossby wave that alternately brings high and low pressures, and various types of waves in plasma. On the other hand, light (electromagnetic waves) and sound waves that are familiar from high school classes are representative of rare types of waves in the sense that they are not dispersive (see the column at the end of this chapter).

When actually finding the dispersion relation, it is more convenient to use Euler's formula

$$e^{i\theta} = \cos\theta + i\sin\theta \tag{3.9}$$

[3]Knowledge of Fourier analysis is essential in the analysis of wave phenomena. The minimum knowledge of Fourier analysis is summarized in Appendix C.

to express the linear sinusoidal wave solution in the complex form as

$$u(x,t) = a\,e^{i(kx-\omega t)} + \text{c.c.}, \quad a = \frac{A}{2}\,e^{i\theta_0}, \tag{3.10}$$

instead of the real expression (3.1). Here "c.c." stands for the complex conjugate of the expression that precedes it.

When the governing equation is a constant coefficient linear PDE like

$$P\left(\frac{\partial}{\partial t}, \frac{\partial}{\partial x}\right)u = 0, \tag{3.11}$$

where P is an arbitrary polynomial, substituting (3.10) causes a replacement

$$\frac{\partial}{\partial t} \longrightarrow -i\omega, \qquad \frac{\partial}{\partial x} \longrightarrow ik, \tag{3.12}$$

and (3.11) becomes

$$P(-i\omega, ik)\,a\,e^{i(kx-\omega t)} = 0. \tag{3.13}$$

Here the condition that the amplitude a is not zero immediately gives the dispersion relation

$$P(-i\omega, ik) = 0. \tag{3.14}$$

The above procedure applies only to the case where the governing equation is a linear PDE with constant coefficients. If the governing equation is nonlinear, a linear sinusoidal wave (3.10) cannot generally be a solution. In the nonlinear case, a procedure called **linearization** is required before obtaining the sinusoidal wave solution and the dispersion relation. Linearization is the transformation of the governing equation into a linear PDE consisting of only first-order terms in $u(x,t)$, by assuming that the wave amplitude is very small. A simple example is given below, so you can see the meaning of this linearization there.

EXAMPLE 1: DISPERSION RELATION OF THE KDV EQUATION

Find the dispersion relation of the KdV equation[4]

$$\frac{\partial u}{\partial t} + c_0\frac{\partial u}{\partial x} + \alpha u\frac{\partial u}{\partial x} + \beta\frac{\partial^3 u}{\partial x^3} = 0. \tag{3.15}$$

[Solution]

Equation (3.15) is nonlinear due to the third term. Under the linear approximation that u is very small, the third term which contains u twice by multiplication is ignored considering

---

[4]The KdV equation is a well-known approximate equation for water waves with long wavelengths and will be dealt with in detail in the next chapter.

it is much smaller than the other terms which contain u only once. This procedure is called the linearization of the equation, giving the linearized KdV equation

$$\frac{\partial u}{\partial t} + c_0 \frac{\partial u}{\partial x} + \beta \frac{\partial^3 u}{\partial x^3} = 0. \tag{3.16}$$

Substituting (3.10) to this yields

$$\left[(-i\omega) + c_0(ik) + \beta(ik)^3\right] a\, e^{i(kx-\omega t)} = 0, \tag{3.17}$$

giving the linear dispersion relation as follows:

$$\omega(k) = c_0\, k - \beta\, k^3. \tag{3.18}$$

This gives a real ω for a real k, so the wave propagates neutrally with neither amplification nor attenuation. Also, the wave velocity $c(k)$ is given by

$$c(k) = \frac{\omega(k)}{k} = c_0 - \beta\, k^2, \tag{3.19}$$

which depends on k, so this is a dispersive wave.   ♣

. . . .

If there are N dependent variables, and therefore the basic equation is a system of simultaneous PDEs, we adopt the form

$$\boldsymbol{u}(x,t) = \begin{pmatrix} a_1 \\ \vdots \\ a_N \end{pmatrix} e^{i(kx-\omega t)} + \text{c.c.} \tag{3.20}$$

as the sinusoidal wave solution. Substituting (3.20) into the linearized system of constant-coefficient linear PDEs causes replacement (3.12), and yields N homogeneous simultaneous linear equations for the amplitude vector $\boldsymbol{a} = {}^t(a_1,\ldots,a_N)$

$$\begin{bmatrix} & & \\ & p_{ij} & \\ & & \end{bmatrix} \begin{pmatrix} a_1 \\ \vdots \\ a_N \end{pmatrix} = \begin{pmatrix} 0 \\ \vdots \\ 0 \end{pmatrix}, \tag{3.21}$$

where $P = (p_{ij})$ is a $N \times N$ matrix determined by the system of equations. Then the condition $\det P \neq 0$ for a nontrivial solution (i.e., $\boldsymbol{a} \neq \boldsymbol{0}$) to exist gives the dispersion relation. If each of the systems of PDE gives the time derivative of each dependent variable such as $\partial u_i/\partial t = \cdots$, the dispersion relation becomes an Nth order algebraic equation for ω, and N different types of waves (wave mode) exist. Then the solution vector $\boldsymbol{a}$ corresponding to each ω teaches us the relationship of the amplitude between dependent variables in that wave mode. Let us try one specific example next.

EXAMPLE 2: THE LINEAR DISPERSION RELATION OF THE LONG WAVE EQUATION

Find the linear dispersion relation of the long water wave equation

$$\frac{\partial \eta}{\partial t} + \frac{\partial [(h + \eta)u]}{\partial x} = 0, \quad \frac{\partial u}{\partial t} + u \frac{\partial u}{\partial x} + g \frac{\partial \eta}{\partial x} = 0, \tag{3.22}$$

mentioned at the end of Chapter 1 and Appendices B and F. Here η is surface displacement, u is horizontal velocity, and h is water depth (constant).

[Solution]

First, linearizing (3.22) around the undisturbed state $(\eta, u) = (0, 0)$,

$$\frac{\partial \eta}{\partial t} + h \frac{\partial u}{\partial x} = 0, \quad \frac{\partial u}{\partial t} + g \frac{\partial \eta}{\partial x} = 0. \tag{3.23}$$

Assuming that

$$\begin{pmatrix} \eta \\ u \end{pmatrix} = \begin{pmatrix} \hat{\eta} \\ \hat{u} \end{pmatrix} e^{i(kx - \omega t)} + \text{c.c.}, \tag{3.24}$$

and substituting into (3.23) yields

$$\begin{pmatrix} -i\omega & hik \\ gik & -i\omega \end{pmatrix} \begin{pmatrix} \hat{\eta} \\ \hat{u} \end{pmatrix} = \begin{pmatrix} 0 \\ 0 \end{pmatrix}. \tag{3.25}$$

The condition that this has a nontrivial solution, that is, the determinant of the coefficient matrix is 0 gives the dispersion relation as follows:

$$\omega^2 = gh k^2 \quad \longrightarrow \quad \omega = \pm\sqrt{gh}\, k, \quad c = \frac{\omega}{k} = \pm\sqrt{gh}. \tag{3.26}$$

It can be seen that, in this system, there are two wave modes for one wavenumber k with the same frequency but opposite directions of propagation. It is a non-dispersive wave because its propagation speed is $\pm\sqrt{gh}$ and does not depend on k. For waves traveling in the positive direction $\omega = \sqrt{gh}\, k$, $\hat{u} = \sqrt{gh} \cdot \hat{\eta}/h$, and for those traveling in the negative direction $\omega = -\sqrt{gh}\, k$, $\hat{u} = -\sqrt{gh} \cdot \hat{\eta}/h$, respectively. From this, we can see that the velocity u at which the water particle actually moves is not the velocity of the wave $c\, (= \sqrt{gh})$ itself but multiplied by the ratio of the wave height η to the depth h.

A tsunami is a typical long water wave, and its property is expected to follow this long wave equation. The propagation velocity of tsunami is given by $\sqrt{gh}$ within the linear approximation. The average water depth in the Pacific Ocean is about 4,000 m, and considering $g = 9.8$ m/s^2, the propagation velocity of a tsunami traveling offshore of the Pacific Ocean is as great as 200 m/s or about 700 km/h. Nevertheless, small fishing boats that encounter the tsunami offshore are not so affected, because the velocity of water movement caused by the tsunami is not the great propagation velocity itself but it is only the value multiplied by the ratio of the amplitude to the depth (if the amplitude is 1 m, 1/4000) to it.   ♣

· ·

In the discussion so far, the space variable is only x, and the wavelike behavior $e^{i(kx-\omega t)}$ is assumed in that direction. However, when the space is multi-dimensional, it often happens that the solution behaves like a wave with respect to one coordinate only, while showing an eigenfunction-like behavior determined from boundary conditions in other directions. The surface water wave, which is often mentioned as an example throughout this book, is a wave of this kind. In the case of such waves, more complex dispersion relations can occur as we will see in the next section.

3.2   LINEAR SINUSOIDAL WAVE SOLUTION OF WATER WAVE

3.2.1   BASIC EQUATIONS OF WATER WAVE

Let us consider waves propagating on the surface of a water layer of depth h. Water is assumed to be inviscid and incompressible fluid, and its velocity field is irrotational and is expressed as the gradient of the velocity potential $\phi(x, z, t)$ which in turn satisfies the Laplace equation.[5] Let $\eta(x, t)$ denote the free surface displacement, x axis be in the propagation direction and the z axis be vertically upward as shown in Fig. 3.2. Then the governing equations and the boundary conditions are given as follows:

$$\phi_{xx} + \phi_{zz} = 0, \qquad\qquad\qquad -h \leq z \leq \eta(x, t) \qquad (3.27\text{a})$$

$$\phi_t + gz + \frac{1}{2}\left(\phi_x^2 + \phi_z^2\right) - \frac{\tau}{\rho}\frac{\eta_{xx}}{\left\{1 + \eta_x^2\right\}^{3/2}} = 0, \qquad z = \eta(x, t) \qquad (3.27\text{b})$$

$$\eta_t + \phi_x\eta_x = \phi_z, \qquad\qquad\qquad z = \eta(x, t) \qquad (3.27\text{c})$$

$$\phi_z = 0, \qquad\qquad\qquad\qquad z = -h. \qquad (3.27\text{d})$$

Here g is gravity acceleration $g = 9.8$ m/s², ρ is water density $\rho = 1000$ kg/m³, and τ is surface tension coefficient, and $\tau = 0.074$ N/m for the interface between water and air at room temperature. The origin of z is taken at the mean water level, so the average of $\eta(x, t)$ with respect to x is zero.

The Laplace equation (3.27a) is a consequence of the assumption of irrotational flow of incompressible fluid, and is a field equation that should hold for the whole region occupied by water. Equation (3.27b) is a dynamic boundary condition that requires the pressure of water to be equal to the atmospheric pressure at the water surface, and (3.27c) and (3.27d) are kinematic boundary conditions requiring that the water does not penetrate the water surface and the bottom, respectively.

[5]The minimum knowledge of fluid mechanics necessary for deriving these basic equations is summarized in Appendix D.

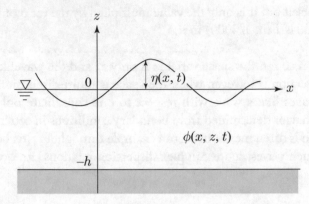

Figure 3.2: A water layer with free surface.

3.2.2   SINUSOIDAL WAVE SOLUTION AND LINEAR DISPERSION RELATION

Since the linear sinusoidal wave solution is the basis for all future discussions, the derivation process is shown in detail below.

If we ignore all the nonlinearities in the basic equations (3.27) to consider waves of small amplitude, we obtain

$$\phi_{xx} + \phi_{zz} = 0, \qquad\qquad -h \leq z \leq 0 \qquad\qquad (3.28\text{a})$$

$$\phi_t + g\eta - \frac{\tau}{\rho}\eta_{xx} = 0, \qquad\qquad z = 0 \qquad\qquad (3.28\text{b})$$

$$\eta_t = \phi_z, \qquad\qquad z = 0 \qquad\qquad (3.28\text{c})$$

$$\phi_z = 0, \qquad\qquad z = -h. \qquad\qquad (3.28\text{d})$$

It should be noted here that the linearization process not only has omitted the nonlinear terms in the equations but also changed the place where the surface boundary conditions are imposed from the actual free surface $z = \eta(x,t)$ deformed by the wave to the static water surface $z = 0$ before being perturbed by the wave. For example, let us consider $\phi(x,z,t)$ contained in (3.27c), which should originally be evaluated at $z = \eta(x,t)$. When expressed by the Taylor expansion around $z = 0$,

$$\phi(x, z = \eta, t) = \phi(x, 0, t) + \eta\frac{\partial\phi(x,0,t)}{\partial z} + \frac{1}{2!}\eta^2\frac{\partial^2\phi(x,0,t)}{\partial z^2} + \cdots. \qquad (3.29)$$

All terms after the second term on the right side are nonlinear terms which contains the wave quantities such as η and ϕ more than twice, and can be ignored in the linear theory that assumes small amplitude. Thus, the boundary conditions at the free surface $z = \eta(x,t)$ can be replaced by those at the undisturbed surface $z = 0$.

Here we assume that the sinusoidal wave solution is of the form:

$$\eta(x,t) = a\, e^{i(kx - \omega t)} + \text{c.c.}, \qquad \phi(x,z,t) = f(z)\, e^{i(kx - \omega t)} + \text{c.c..} \qquad (3.30)$$

Since we do not know the z dependence of ϕ, so for the time being we leave it as $f(z)$. Substituting this into (3.28a) gives

$$\frac{d^2 f}{dz^2} - k^2 f = 0, \qquad (3.31)$$

showing that $f(z)$ should be a linear combination of e^{kz} and e^{-kz}. In addition to this, considering the condition $\left.\frac{df(z)}{dz}\right|_{z=-h} = 0$ required by (3.28d), $f(z)$ is obtained as follows[6]:

$$f(z) = C \cosh[k(z + h)]. \quad (C \text{ arbitrary complex constant}) \qquad (3.32)$$

Substituting this result into (3.28c) gives

$$-i\omega a = \left.\frac{df(z)}{dz}\right|_{z=0} = k\,C \sinh kh, \quad \longrightarrow \quad C = \frac{-i\omega a}{k \sinh kh}. \qquad (3.33)$$

And substituting these into (3.28b) yields

$$-i\omega f(0) + ga - \frac{\tau}{\rho}(ik)^2 a = 0,$$

$$\longrightarrow \quad -i\omega \left(\frac{-i\omega a}{k \sinh kh}\right) \cosh kh + \left(g + \frac{\tau}{\rho}k^2\right) a = 0, \qquad (3.34)$$

from which the linear dispersion relation

$$\omega^2(k) = \left(gk + \frac{\tau}{\rho}k^3\right) \tanh kh, \qquad (3.35)$$

and the sinusoidal wave solution

$$\eta(x,t) = A \cos(kx - \omega t + \theta_0), \qquad (3.36a)$$

$$\phi(x,z,t) = \frac{\omega A}{k} \frac{\cosh[k(z + h)]}{\sinh kh} \sin(kx - \omega t + \theta_0) \qquad (3.36b)$$

are obtained. It can be seen that the phase velocity $c(k)$ is given by

$$c(k) = \frac{\omega}{k} = \pm\sqrt{\left(\frac{g}{k} + \frac{\tau}{\rho}k\right) \tanh kh}. \qquad (3.37)$$

---

[6]$\cosh x$, $\sinh x$ are defined by $\cosh x = (e^x + e^{-x})/2$, $\sinh x = (e^x - e^{-x})/2$, respectively, and along with $\tanh x$ mentioned previously, are called the hyperbolic functions.

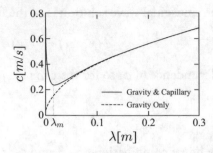

Figure 3.3: Phase velocity of water wave. Solid: capillary gravity wave; dotted: gravity wave (water depth is infinite).

Figure 3.3 shows the phase velocity (3.37) of the surface wave as a function of wavelength $\lambda (= 2\pi/k)$, when the water depth h is infinite. The solid line in the figure is (3.37) itself, and the dotted line is the value when ignoring the effect of surface tension with $\tau = 0$. The terms g and τ in parentheses in (3.35) and (3.37) indicate the contribution of gravity and surface tension as the restoring force, respectively. If we let k_m be the wavenumber where the contribution of gravity and surface tension become equal,

$$gk_m = \frac{\tau}{\rho}k_m^3 \quad \longrightarrow \quad k_m = \sqrt{\frac{\rho g}{\tau}}. \tag{3.38}$$

In the case of the interface between water and air, $k_m = 3.6$ rad/cm, and the corresponding wavelength is about $\lambda_m = 2\pi/k_m = 1.7$ cm. Equation (3.35) indicates that the contribution of gravity to ω^2 is proportional k, while that of surface tension is proportional to k^3. Therefore, the gravity becomes dominant as the wavelength becomes longer ($k < k_m$) and the surface tension becomes dominant as the wavelength becomes shorter ($k > k_m$). In the case of water and air, for example, at $\lambda = 10$ cm, the contribution from surface tension already decreases to less than 3% of the contribution from gravity. In ocean waves, most of the energy is possessed by waves with wavelengths of several tens of meters or more, and for such waves, the effect of surface tension can be ignored almost completely. From this point onward, we will ignore surface tension and consider water waves taking into account only of gravity as the restoring force, unless stated otherwise. Such wave is called **gravity wave**.[7]

As Fig. 3.3 shows, when the water depth is infinite, the phase velocity of water wave takes a minimum value of 0.23 m/s at the wavelength λ_m, and it becomes faster if the wavelength is longer or shorter than λ_m. The dotted line in the figure is the phase velocity of the gravity wave neglecting the effect of surface tension. It can be confirmed that as the wavelength becomes longer than λ_m, the phase velocity of water wave considering both gravity and surface tension

[7]Be careful not to confuse it with the "gravitational wave" that is the wave of distortion of space-time that Einstein predicted its existence in the theory of general relativity.

rapidly approaches to that of the gravity wave, that is, the effect of surface tension becomes negligible.

However, the micro-scale wave breaking, for which the surface tension plays effectively, may play an important role in the exchange of materials (such as CO_2 and aerosols) and energy between the atmosphere and the ocean as a total amount. In remote sensing of sea waves by satellites, etc., electromagnetic waves called "microwaves" of several centimeter in wavelength are used. What directly participates in the reflection of these electromagnetic waves through Bragg scattering is water waves of a few centimeter wavelength, that is, waves for which surface tension works effectively. For reasons as above and others, it should be noted that the study of wave phenomena on a short scale that are directly affected by surface tension has no less importance than the study of gravity waves. Also, if it is not limited to sea surface waves, there are many situations where surface tension plays a decisive role, for example, flow phenomena in a microgravity environment such as the International Space Station (for example, how to make a more homogeneous silicon wafer not affected by gravity) or flow phenomena on a microscale such as those in capillaries of human body or micromachines.

The dispersion relation and phase velocity of gravity wave are given by

$$\omega(k) = \sqrt{gk \tanh kh}, \qquad c(k) = \frac{\omega}{k} = \sqrt{\frac{g}{k} \tanh kh}, \qquad (3.39)$$

by putting $\tau = 0$ in (3.35). The sinusoidal wave solution (3.36a) and (3.36b) can be used as they are with understanding that ω is given by (3.39) instead of (3.35). The relationship between c and λ ($= 2\pi/k$) is shown in Fig. 3.4. Here, c and λ are normalized by the linear long-wave phase speed $\sqrt{gh}$ and the water depth h, respectively. As can be seen from the figure, when the water depth is fixed, the phase velocity of the gravity wave increases monotonically as the wavelength becomes longer, and it takes the maximum value $\sqrt{gh}$ in the limit $\lambda \to \infty$.

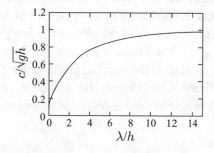

Figure 3.4: Phase velocity of gravity wave as a function of wavelength.

3.2.3   DEEP WATER (OR SHORT WAVE) LIMIT

According to (3.39), the wave velocity $c(k)$ is given by $\sqrt{g/k}$ in the limit $kh \to \infty$ where the water depth is very deep compared to the wavelength. Numerically, $\sqrt{\tanh kh}$ takes values between 0.97 and 1 for $h > 1.75/k \approx 0.28\lambda$. This implies that c given by (3.39) differs from the limiting value $\sqrt{g/k}$ by less than 3% when the water depth is more than 30% of the wavelength. Thus, at least with regard to the wave velocity, if the water depth is more than only 30% of the wavelength, it will be almost the same as when the water depth is infinitely deep. If we rewrite the dispersion relation of deep water gravity wave $\omega^2 = gk$ to the relations between λ [m], c [m/s], T [s], we obtain

$$\lambda = \frac{g}{2\pi}T^2 = 1.56T^2, \qquad c = \frac{\lambda}{T} = 1.56T. \qquad (3.40)$$

These equations are useful for estimating the wavelength and wave speed from the wave period. For example, the wavelength and the wave speed of waves with a period of 8 s in the open ocean can be immediately estimated to be about 100 m and 12.5 m/s, respectively.

3.2.4   SHALLOW WATER (OR LONG WAVE) LIMIT

In the shallow water limit $kh \to 0$, $c(k) \to \sqrt{gh}$ since $\tanh kh \to kh$.[8] That is, when the wavelength becomes very long compared to the water depth, the wave velocity becomes gradually independent of the wavelength, and the dispersion disappears. $\sqrt{\tanh kh}/\sqrt{kh}$ takes numerical values between 0.97 and 1 when $kh < 0.43$. This means that if $\lambda > 15h$, the difference with the phase speed of the truely long wave $\sqrt{gh}$ is less than 3%. However, remembering that the water depth can be regarded virtually infinite if the water depth is more than 30% of the wavelength, the condition for being regarded as a shallow water is more stringent that it needs to be less than 7% of the wave length.

The average water depth in the Pacific Ocean is about 4000 m. The condition $kh < 0.43$ for long waves for $h = 4000$ m corresponds to about 58 km or more in wavelength and about 300 s in period. A tsunami with a period of about 10 min is completely within the range that can be treated as a long wave. That is, even the ocean with a depth of 4000 m is a very shallow water from the viewpoint of the tsunami. The wave velocity of the long wave is given by $\sqrt{gh}$, so the velocity of the tsunami crossing the Pacific Ocean is about 200 m/s, or 700 km/h or more. The tsunami generated by the earthquake off the coast of Chile in South America is known to reach coasts of Japan in almost a full day. Considering that the distance between Japan and Chile is approximately 17,000 km, this required time is quite consistent with the speed of a long wave.

3.2.5   REFRACTION

When the ocean waves pass through a region of decreasing water depth and gradually approach the coastline from the offshore, their frequency ω tends to be kept approximately constant.[9] According to the dispersion relation (3.39) of the gravity wave in finite water depth, the wavelength λ and the phase velocity c become increasing functions of the water depth h when ω is constant.

---

[8]The Taylor expansion of $\tanh x$ around $x = 0$ is $\tanh x = x - x^3/3 + 2x^5/15 + \cdots$, so $\tanh x \approx x$ as $x \ll 1$.
[9]This property is addressed in Section 6.1 in relation to the group velocity.

Therefore, the wavelength and velocity decrease as the wave enters from the offshore to the shallow region. For example, in the case of a wave with a period of 8 s, $\lambda = 99.8$ m, and $c = 12.5$ m/s in deep water, but when it reaches a point with $h = 5$ m, $\lambda = 53.14$ m, $c = 6.63$ m/s, and both decrease to about 1/2 of the offshore values. (See Example 3.)

When standing on a beach, the waves always seem to be coming straight from the offshore toward us. If the beach is curved, the normal vector of the shoreline (i.e., the vector pointing straight to the offshore) points in different directions at different points on the shoreline. Nevertheless, no matter where we are on the beach, the waves come from the direction of the normal vector, with the wave peaks being almost parallel to the beach. The waves coming to the beach when there is no wind around the beach are the waves that were generated by a storm at a distant place several days ago, which are called the **swell**. Therefore, if you observe the swell offshore from the beach, the direction in which the swell comes is the direction in which the storm that generated them was. Bearing this in mind, it's a little strange that waves come in at a right angle to the shoreline no matter where we are on a curved beach.

The phenomenon that the wave direction tends to align to the depth contour near the beach in this way is due to the nature of the gravity wave that the wave velocity decreases as the water depth decreases as seen above. As shown in Fig. 3.5, when the wave is incident obliquely to the coast, the speed of the part that first enters the shallow area is slower, but the part that is still in the deep area continues to propagate at a high speed. As a result, the crest always rotates in the direction toward the coast. When a beam of light is incident on the water surface from the air, the light bends toward the water in which the speed of light is slower than in the air. This is the phenomenon of **refraction** and is familiar to anyone. Just the same thing is happening for water waves when approaching the coast (but not only once but continuously). Figure 3.6 shows an image of the change of the direction (refraction) of waves. As can be seen from the figure, the energy of waves tends to concentrate at the cape, and it conversely tends to be dispersed in the bay area. When driving a car along the coastline, it is often observed that waves are broken at the tip of the cape and white waves are standing. This is because the energy of the waves tends to be concentrated there and the wave height becomes high. On the other hand, the bathing beaches where we want the waves to be calm are created in the bay area where the energy of the waves is dispersed and the wave height is low.

EXAMPLE 3: FINDING THE WAVELENGTH FROM THE PERIOD

Find the wavelength and wave velocity at a depth of 5 m of the gravity wave with a period of 8 s.

[Solution]

It is easy to find the frequency for a given wavenumber using the dispersion relation (3.39). However, given the period or frequency, it is not so easy to find the corresponding wavenumber or wavelength. Here, let's find an approximate solution numerically using Newton's method

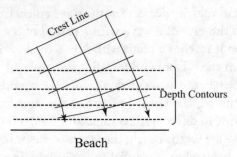

Figure 3.5: Alignment of crest line to the depth contour.

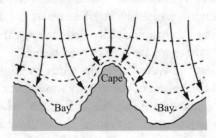

Figure 3.6: Wave refraction around cape and bay.

(Newton–Raphson method), which is a method to find the solution of a nonlinear equation numerically.

First, let's review Newton's method briefly. Consider finding the solution of the nonlinear equation $f(x) = 0$ by an iterative method. Let the nth approximate solution of the iteration be represented by x_n. Since x_n has not yet reached the true solution, $f(x_n) \neq 0$. By Taylor expanding $f(x)$ around x_n,

$$f(x) = f(x_n) + f'(x_n)(x - x_n) + O(\Delta x^2). \tag{3.41}$$

If we ignore the second- or higher-order terms of the expansion here and request that the next approximate solution x_{n+1} to satisfy $f(x_{n+1}) = 0$, then we obtain

$$0 = f(x_{n+1}) = f(x_n) + f'(x_n)(x_{n+1} - x_n) \implies x_{n+1} = x_n - \frac{f(x_n)}{f'(x_n)}. \tag{3.42}$$

Since we have ignored the second- and higher-order terms of the expansion, this x_{n+1} cannot be a true solution, but it can be an improved approximate solution than x_n. The numerical method for finding the solution of the nonlinear equation $f(x) = 0$ by the recursion (3.42) is known as **Newton's method**.

In the case of this example, let

$$f(k) := gk \tanh(kh) - \omega^2, \quad g = 9.8, \ h = 5.0, \ \omega = \frac{2\pi}{T} = \frac{\pi}{4}; \tag{3.43}$$

then the recursion formula of the Newton's method becomes

$$k_{n+1} = k_n - \frac{gk \tanh(kh) - \omega^2}{g \tanh(kh) + gkh[1 - \tanh^2(kh)]}. \tag{3.44}$$

If, for example, a value $k_0 = \omega^2/g$ corresponding to a period of 8 s when $h \to \infty$ is adopted as the starting value k_0 and (3.44) is repeated, it converges to $k = 0.118435696\cdots$ in about 4 or 5 iterations. From this, the wavelength λ and the wave velocity of the wave with period of 8 s at a water depth of 5 m are given as

$$\lambda = 2\pi/k \approx 53.1 \,\text{m}, \quad c = \omega/k \approx 6.63 \,\text{m/s}. \tag{3.45}$$

♣

. . . .

3.2.6   MOTION OF WATER PARTICLE

According to the sinusoidal wave solution for gravity waves,

$$\eta(x,t) = A \cos(kx - \omega t), \tag{3.46a}$$

$$\phi(x,z,t) = \frac{\omega A}{k} \frac{\cosh[k(z+h)]}{\sinh kh} \sin(kx - \omega t), \tag{3.46b}$$

$$\omega = \sqrt{gk \tanh kh}, \tag{3.46c}$$

the motion of a certain water particle, i.e., the temporal change of its coordinates $[x(t), z(t)]$ is described by

$$\frac{dx}{dt} = u(x,z,t) = \frac{\partial\phi}{\partial x} = \omega A \frac{\cosh[k(z+h)]}{\sinh kh} \cos(kx - \omega t), \tag{3.47a}$$

$$\frac{dz}{dt} = w(x,z,t) = \frac{\partial\phi}{\partial z} = \omega A \frac{\sinh[k(z+h)]}{\sinh kh} \sin(kx - \omega t). \tag{3.47b}$$

Since the unknown functions $x(t), z(t)$ are included in the trigonometric functions on the right sides, $x(t), z(t)$ cannot be solved explicitly as it is. However, in the linear wave theory that the amplitude is infinitesimal, $x(t)$ and $z(t)$ on the right side can be replaced by the positions x_0, z_0 when the wave amplitude is zero. So,

$$\frac{dx}{dt} \approx \omega A \frac{\cosh[k(z_0+h)]}{\sinh kh} \cos(kx_0 - \omega t), \tag{3.48a}$$

$$\frac{dz}{dt} \approx \omega A \frac{\sinh[k(z_0+h)]}{\sinh kh} \sin(kx_0 - \omega t). \tag{3.48b}$$

This gives

$$x - x_0 = -R_x \sin(kx_0 - \omega t), \quad R_x = A\frac{\cosh[k(z_0 + h)]}{\sinh kh}, \tag{3.49a}$$

$$z - z_0 = R_z \cos(kx_0 - \omega t), \quad R_z = A\frac{\sinh[k(z_0 + h)]}{\sinh kh}. \tag{3.49b}$$

If t is eliminated from this equation, it can be seen that the water particle draws an elliptical orbit

$$\frac{(x - x_0)^2}{R_x^2} + \frac{(z - z_0)^2}{R_z^2} = 1, \tag{3.50}$$

with a major (horizontal) axis R_x and minor (vertical) axis R_z, as shown in Fig. 3.7.

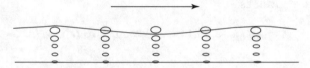

Figure 3.7: Elliptic orbit of water particle.

From this result we can see the properties of the wave motion as follows.

1. At $z_0 = 0$, that is, on the mean water surface, the length of the minor axis R_z coincides with the amplitude A of the surface displacement η.

2. In the case of virtually deep water waves ($kh \gg 1$), except for the vicinity of the bottom, $R_x \approx R_z \approx A\,e^{kz_0}$, so the motion of water particle follows a circular orbit.

   Also, its radius decreases exponentially as it goes away from the water surface. Specifically, the wave amplitude decreases by $e^{-2\pi} \approx 1/535$ every time the distance from the surface increases by one wavelength. When swimming in a swimming pool or the sea, many readers may have experienced that only a little dive leads to a rapid decrease of the wave motion even if there is a large wave on the surface.

3. In the case of virtually shallow water ($kh \ll 1$),

$$R_z/R_x = \tanh[k(z_0 + h)] \to 0, \qquad (kh \to 0) \tag{3.51}$$

showing that the ellipse becomes a horizontally crushed shape, and the water particle move almost in the horizontal direction only.

It may seem a little strange that the water particles only move in an elliptical motion and almost stays in the same place forever, but the waves go on and on. Figure 3.8 shows the relation between the wave propagation and the circular motion of water particles in the case of deep water wave. You can see how the water particles transmit the (phase of) waves by performing circular motion with slight phase differences.

Figure 3.8: Wave propagation and circular motion of water particles.

Note that, under the linear approximation assuming infinitesimal amplitude adopted here, the orbit of each water particle is a closed ellipse, and therefore the wave does not produce mass flux. However, this does not hold when going to the nonlinear theory that takes into account the finiteness of wave amplitude. When the water depth is deep, the water particles in the water perform circular motion, but the radius of this circular motion becomes smaller exponentially as the distance from the surface gets larger. When the amplitude of the wave increases, the part of the circular motion of a water particle that travels in the same direction of the wave passes a slightly shallower part, so the radius of movement at that time is a little larger than the value at the center, while the part that moves backward the particle moves in a little deeper part, so the radius of movement is a little smaller. From this difference in radius and velocity of water particle in different part of the circular motion, the trajectory of water particle changes from a closed circular trajectory to a trajectory that shifts slightly forward with each rotation, resulting in a substantial mass flux that is proportional to the square of the wave amplitude. This is called the **Stokes drift**.

3.2.7   DISPERSION RELATION BY DIMENSIONAL ANALYSIS

In Section 3.2, the linear dispersion relation is derived by solving the system of basic equations of water waves. However, using the method of "dimensional analysis" introduced below, this dispersion relation can be found much more easily.

Before doing that, let us first start with a simpler problem of finding the period T [s] of a single pendulum as shown in Fig. 3.9 consisting of a weight of mass m [kg] and a string of length l [m]. If we start from the equation of motion seriously, it would become as follows. Newton's second law of motion for the tangential direction gives

$$ml\ddot{\theta} = -mg\sin\theta. \tag{3.52}$$

If the amplitude is small, $\sin\theta$ can be approximated by θ, then

$$ml\ddot{\theta} = -mg\theta \longrightarrow \ddot{\theta} = -\omega^2\theta, \quad \omega \equiv \sqrt{g/l}, \tag{3.53}$$

and the general solution for this is given by

$$\theta(t) = A\cos(\omega t + \theta_0), \quad A, \theta_0 : \text{const.} \tag{3.54}$$

This solution shows that the period T of the pendulum is given by $T = 2\pi/\omega = 2\pi\sqrt{l/g}$.

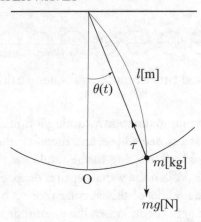

Figure 3.9: A single pendulum.

However, there is a totally different way as follows to find this period T which can be carried out by an elementary school student who does not know Newton's law or differential equations, or do not even know the differentiation. First, the factors that may affect the period T [s] are considered to be the length l [m] of the pendulum, mass of the weight m [kg], and the gravitational acceleration g [m/s^2]. Therefore, T should be represented by some combination of l, m, and g. Since the unit of T is [s], we need to find a combination of l, m, and g which has the same unit [s]. Then there is only $\sqrt{l/g}$ that meets this requirement. From this alone, it can be inferred that T should be proportional to $\sqrt{l/g}$, i.e., $T = \alpha\sqrt{l/g}$, with α being a dimensionless number.

A method that considers the relationship between physical quantities by relying only on units of the quantities (more precisely, dimensions) in this way is called the **dimensional analysis**. Since it is not possible to determine the dimensionless coefficient α in dimensional analysis, it is not possible to know the value of T itself. But important properties of T, such as it is proportional to $\sqrt{l/g}$ and it does not depend of the mass m, can be known without any complicated calculations.

Let's apply this method of dimensional analysis to the wave velocity of the gravity wave. Factors that may influence the wave velocity c [m/s] include the wavelength λ [m], the density ρ [kg/m^3] of water (or the liquid, in general), gravitational acceleration g [m/s^2], and the water depth h [m]. First, let us consider the deep water wave limit $h/\lambda \to \infty$. The movement of water generated by the wave decays exponentially rapidly as it moves away from the free surface as seen above. Therefore, when the water depth is deep enough, the influence of the surface wave does not reach the bottom of water, and as a result, the surface wave does not recognize where the bottom is. In such a situation, the water depth h should not affect c, so c should be able to be written by a combination of λ, ρ, and g. Considering the combination of these three that has the same unit [m/s] as c, there is only $\sqrt{g\lambda}$. If this fact is written as the relation between ω and

k,

$$c = \beta \sqrt{g\lambda} = \tilde{\beta}\sqrt{g/k} \quad \longrightarrow \quad \omega = ck = \tilde{\beta}\sqrt{gk}. \tag{3.55}$$

Thus, we can get the same dispersion relation of deep water wave as that derived from the system of basic equations, except for the undetermined dimensionless constant β.

Next, let us consider the shallow water limit $\lambda/h \to \infty$. What is important in this limit is not the specific value of λ but only the fact that λ is much larger than h. Hence, λ seems not to appear in the expression for c. If this guess is correct, c can be written by the combination of h, ρ, and g, and a relationship as

$$c = \gamma\sqrt{gh}, \quad \omega = \gamma\sqrt{gh}\,k \tag{3.56}$$

is obtained by searching for a combination whose unit is [m/s]. This is consistent with the shallow water limit of the linear dispersion relation derived from the basic system of equations again, except for the dimensionless constant γ.

These dimension-based considerations are often used throughout this book. The basic theorems and application examples of dimensional analysis are summarized in Appendix E.

3.3   WAVE ENERGY AND ITS PROPAGATION VELOCITY

In this section, we will consider the energy associated with a sinusoidal gravity wave and its propagation velocity.

3.3.1   KINETIC ENERGY AND POTENTIAL ENERGY

Let V be the region occupied by water which corresponds to one wavelength of the wave, and S be the part of the boundary of V that corresponds to the free surface. Then the total amount of kinetic energy in V is given by

$$\iiint_V \frac{1}{2}\rho v^2 \, dV = \frac{1}{2}\rho \iiint_V (\nabla\phi)^2 \, dV = \frac{1}{2}\rho \iiint_V \nabla \cdot (\phi\nabla\phi) \, dV = \frac{1}{2}\rho \iint_S \phi \frac{\partial\phi}{\partial n} \, dS. \tag{3.57}$$

This is an exact expression that does not require the approximation of small amplitude. In the first equal sign, it is assumed that the water is incompressible and the density ρ can be treated as a constant, and moreover the motion is irrotational and the velocity vector v is given by the velocity potential ϕ as $v = \nabla\phi$. In the second equal sign it is used that ϕ satisfies the Laplace equation (3.27a), and in the third equal sign the Gauss's divergence theorem, the periodicity of the wave in the horizontal direction and the bottom boundary condition $\partial\phi/\partial n = 0$ are used. Here $\partial\phi/\partial n$ represents the derivative of ϕ in the outward normal direction.

Under the linear approximation, the surface element dS of the free surface can be replaced by the surface element of the flat surface before being deformed by the wave, and $\partial\phi/\partial n$ can also be replace by a vertical derivative. Then the kinetic equation K per unit horizontal area of a linear wave is given by

$$K = \frac{1}{2}\rho \left(\phi \frac{\partial \phi}{\partial z} \right)_{z=0}.$$

(3.58)

Also, this can be rewritten as

$$K = \frac{1}{2}\rho \frac{1}{k \tanh kh} \left(\frac{\partial \eta}{\partial t} \right)^2$$

(3.59)

by using the linearized kinetic boundary condition at the free surface

$$\left. \frac{\partial \phi}{\partial z} \right|_{z=0} = \frac{\partial \eta}{\partial t}$$

(3.60)

and the relation derived from the linear sinusoidal wave solution (3.36) as follows:

$$\phi|_{z=0} = \frac{\omega A}{k} \frac{1}{\tanh kh} \sin(kx - \omega t) = \frac{1}{k \tanh kh} \frac{\partial \eta}{\partial t}.$$

(3.61)

The average over one period $\overline{K}$ of the kinetic energy per unit horizontal area is given by

$$\overline{K} = \frac{1}{2}\rho \frac{1}{k \tanh kh} \overline{\left(\frac{\partial \eta}{\partial t} \right)^2} = \frac{1}{4}\rho \frac{\omega^2}{k \tanh kh} A^2 = \frac{1}{4}\rho g A^2,$$

(3.62)

where (3.39) has been used.[10]

On the other hand, the potential energy P per unit horizontal area, relative to the undisturbed free surface $\eta = 0$ is given by

$$P = \int_0^\eta \rho g z \, dz = \frac{1}{2}\rho g \eta^2,$$

(3.63)

hence the average of the potential energy $\overline{P}$ per unit horizontal area is given by

$$\overline{P} = \frac{1}{2}\rho g \overline{\eta^2} = \frac{1}{4}\rho g A^2.$$

(3.64)

After all, the average energy $\overline{E}$ of wave energy per unit horizontal area is given by

$$\overline{E} = \overline{K} + \overline{P} = \frac{1}{2}\rho g A^2.$$

(3.65)

[10]For example, the average over one period of $\cos^2(kx - \omega t)$ is given by

$$\overline{\cos^2(kx - \omega t)} \equiv \frac{1}{T} \int_0^T \cos^2(kx - \omega t) \, dt = \frac{1}{2\pi} \int_0^{2\pi} \cos^2 \theta \, d\theta = \frac{1}{2},$$

and the same applies to $\sin^2(kx - \omega t)$.

It can be seen that the equipartition between kinetic and potential energies is established in terms of their average over one wave period.

Looking at (3.63), the potential energy is proportional to η^2, and is positive not only at the crest ($\eta > 0$) but also at the trough ($\eta < 0$) of the wave. It may seem a bit strange that the potential energy is positive even at the trough where the water level has been lowered. However, this is natural once we remember that it is water that can have the potential energy. The water which is above the reference level, i.e., the flat surface $z = 0$, has positive potential energy, while the water below $z = 0$ has negative potential energy. At the crest of the wave, since the water which is above the reference level and hence has positive potential energy is newly added by the wave, the potential energy there is naturally positive. On the other hand, at the trough of the wave, the water which used to be there before the wave occurs has been removed. The water below the reference level has negative potential energy. Since the water with negative potential energy has been removed, the potential energy has also increased to become positive at the trough, too.

3.3.2   ENERGETIC CONSIDERATION ON THE DISPERSIVITY OF WATER WAVES

An interpretation can also be given to the dispersive character of water waves from the viewpoint of energy. Here, we introduce it following the argument by Lighthill [5]. First, let's start with a review of the basic oscillatory motion of a system of mass m and spring constant k.[11] Denoting the displacement as $x(t)$, the equation of motion is $m\ddot{x} = -kx$. The potential energy of this motion is $\frac{1}{2}kx^2$, and its average per period is $\frac{1}{2}k\overline{x^2}$. The kinetic energy is $\frac{1}{2}m\dot{x}^2$, and its average per period is given by $\frac{1}{2}m\omega^2\overline{x^2}$. The general solution of this motion is given by $x(t) = A\cos(\omega t + \theta_0)$, $\omega = \sqrt{k/m}$, and it can be understood that ω is determined so that the general property of small oscillation around an equilibrium that "the average potential energy per period and the average kinetic energy per period is equal" holds. This also means that ω^2 is determined as the ratio of the "stiffness" k, which is the coefficient of $\frac{1}{2}x^2$ in the potential energy, and the "inertia" m, which is the coefficient of $\frac{1}{2}\dot{x}^2$ in the kinetic energy.

Let's apply this interpretation of ω^2 to the water wave. The expression (3.63) of the potential energy P and the expression (3.59) of kinetic energy K for small amplitude water waves have forms that are proportional to η^2 and $\dot{\eta}^2$, respectively. If η is regarded as the generalized coordinate, the ratio of the respective coefficients, the "stiffness" ρg and the "inertia" $\rho/k \tanh kh$, gives the correct linear dispersion relation $\omega^2 = gk \tanh kh$. As a result, even in the case of water waves, the equipartition $\overline{K} = \overline{P}$ is established as in the case of the simple vibration of a spring.

Thus, if we adopt the vertical displacement η as the generalized coordinate, the "stiffness" ρg does not depend on wavelength, but the "inertia" $\rho/k \tanh kh$ changes depending on wavelength. This is because the degree of penetration of motion in the direction toward the bottom changes depending on the wavelength, and also because the ratio of kinetic energy in the hori-

---

[11]For a general theory of small oscillation around an equilibrium, see, for example, Chapter 6 of Goldstein [2].

zontal direction to that in the vertical direction changes depending on the wavelength. It can be said that this wavelength dependence of "inertia" produces the wavelength dependence of the frequency, that is, the dispersion of water waves.

3.3.3 ENERGY FLUX AND VELOCITY OF ENERGY PROPAGATION

Next, let us consider the energy flux and energy propagation velocity associated with the linear sinusoidal wave (3.46). The energy flux, that is, the flow of energy F_x in the positive x direction per unit time, which crosses a vertical cross section perpendicular to the wave propagation directions, is composed of two parts: (1) by a portion of water with energy crossing the section; and (2) by the excess pressure generated by the wave doing work.

If the wave amplitude is denoted as a, the energy density is $O(a^2)$ and the flow velocity is $O(a)$, so the part (1) is a quantity of $O(a^3)$. On the other hand, in the part (2), the excess pressure by the wave is $O(a)$, and the displacement per unit time, that is, the flow velocity is $O(a)$, so the magnitude of its contribution is of $O(a^2)$. Therefore, in linear theory where a is very small, the part (1) can be ignored compared to the part (2).

Linearizing the Euler equation[12] which is an equation of motion for an inviscid fluid, and write its x component we get

$$\frac{\partial u}{\partial t} = -\frac{1}{\rho}\frac{\partial p}{\partial x}. \tag{3.66}$$

By using $u = \partial\phi/\partial x$ here, we obtain

$$p = -\rho\frac{\partial\phi}{\partial t} = \rho c\frac{\partial\phi}{\partial x}. \tag{3.67}$$

Here, for the second equal sign, we have used the fact that $\frac{\partial}{\partial t} = -c\frac{\partial}{\partial x}$ holds because the sinusoidal wave translates at a speed c. Therefore, for a linear sinusoidal wave, the energy flux F_x is given by

$$F_x = \int_{-h}^{0} pu\, dz = \int_{-h}^{0} \rho c\frac{\partial\phi}{\partial x}\frac{\partial\phi}{\partial x}\, dz = 2c\int_{-h}^{0}\frac{1}{2}\rho\left(\frac{\partial\phi}{\partial x}\right)^2 dz = 2c\, K_x, \tag{3.68}$$

where K_x $(= \int_{-h}^{0}\frac{1}{2}\rho(\phi_x)^2 dz)$ is the horizontal kinetic energy per unit surface area.

The propagation velocity U of energy should be defined as the ratio of the average energy flux $\overline{F_x}$ to the average energy density $\overline{E}$ $(= \overline{K} + \overline{P})$. So we have

$$U = \frac{\overline{F_x}}{\overline{K} + \overline{P}} = \frac{2c\overline{K_x}}{2\overline{K}} = c\frac{\overline{K_x}}{\overline{K}} \quad\longrightarrow\quad \frac{U}{c} = \frac{\overline{K_x}}{\overline{K}}, \tag{3.69}$$

that is,

$$\frac{\text{Energy propagation velocity}}{\text{wave velocity (phase velocity)}} = \frac{\text{average horizontal kinetic energy}}{\text{average total kinetic energy}}. \tag{3.70}$$

---

[12] For Euler equation in fluid mechanics, see Appendix D.

If this is specifically calculated using the sinusoidal solution (3.46),

$$\frac{U}{c} = \frac{\overline{K_x}}{\overline{K}} = \frac{\int_{-h}^{0} \overline{u^2} \, dz}{\int_{-h}^{0} \overline{(u^2 + w^2)} \, dz} = \frac{\int_{-h}^{0} \cosh^2[k(z+h)] \, dz}{\int_{-h}^{0} \{\cosh^2[k(z+h)] + \sinh^2[k(z+h)]\} \, dz}$$

$$= \frac{\int_{-h}^{0} \frac{1}{2} \{1 + \cosh[2k(z+h)]\} \, dz}{\int_{-h}^{0} \cosh[2k(z+h)] \, dz} = \frac{1}{2} \left\{1 + \frac{2kh}{\sinh 2kh}\right\}. \tag{3.71}$$

Then,

$$U = \frac{1}{2} \left\{1 + \frac{2kh}{\sinh 2kh}\right\} c = \begin{cases} c, & \text{shallow water limit } (kh \to 0), \\ \frac{1}{2}c, & \text{deep water limit } (kh \to \infty). \end{cases} \tag{3.72}$$

The last result concerning the ratio of U and c in the two limiting situations can be derived directly from (3.69) by considering that, in the shallow water limit $kh \to 0$, the motion of water particle is almost horizontal, so $\overline{K_x} \gg \overline{K_z}$, i.e., $\overline{K} \approx \overline{K_x}$, and in the deep water limit $kh \to \infty$, on the other hand, the motion of water particle is circular and isotropic, so $\overline{K_x} = \overline{K_z}$, i.e., $\overline{K} = 2\overline{K_x}$.

Many readers may have already learned somewhere that "energy propagates with group velocity $v_g = \frac{d\omega}{dk}$." However, the velocity which is appropriate to be called the "energy propagation velocity" should be defined as the ratio of energy flux to energy density as above. Let's confirm by the following example whether the energy propagation velocity U defined in this way and the group velocity v_g defined by the k derivative of $\omega(k)$ really agree.

EXAMPLE 4: EQUALITY OF ENERGY PROPAGATION VELOCITY AND GROUP VELOCITY

Calculate the group velocity $v_g = \frac{d\omega}{dk}$ and confirm that it is equal to U of (3.71).

[Solution]

$$\omega^2 = gk \tanh kh \quad \longrightarrow \quad 2\omega \frac{d\omega}{dk} = g \tanh kh + \frac{gkh}{\cosh^2 kh}. \tag{3.73}$$

Therefore,

$$\frac{d\omega}{dk} = \frac{1}{2\omega} \frac{\omega^2}{k} + \frac{1}{2\omega} \frac{gkh}{\cosh^2 kh} = \frac{1}{2}c + \frac{1}{2}c \frac{gk^2h}{\omega^2 \cosh^2 kh} = \frac{1}{2} \left\{1 + \frac{2kh}{\sinh 2kh}\right\} c, \tag{3.74}$$

and it is true that v_g is equal to U.   ♣

. . . .

But as we saw above, the derivation process of the two velocities are completely differ-ent. The derivation of U is by integration with respect to z of the sinusoidal solution, while the derivation of v_g is by differentiation of $\omega(k)$ with respect to k. Also, as you see from the above derivation process, the energy propagation velocity U is a quantity that can be considered for a completely monochromatic sinusoidal wave, while the group velocity v_g is given by a k-derivative, that is, a quantity that cannot be considered without being aware of the difference from another sinusoidal wave with a slightly different wavenumber. Considering this way, there may be many readers who wonders why the energy propagation velocity U defined by the "ratio of energy flux to energy density" always coincides with the group velocity v_g defined by the "k-derivative of ω, where there is the necessity for the two to coincide.[13] We will not pursue this question further here. For readers who are interested, I recommend taking a look at Section 3.4 of [5]. The group velocity v_g will be discussed again in Chapter 6 of this book in the context of modulation of wavetrains.

3.4 EXTENSION OF LINEAR SOLUTION TO NONLINEAR SOLUTION

3.4.1 CRITERIA FOR VALIDITY OF LINEAR APPROXIMATION

The system of basic equations (3.27) of surface water waves is nonlinear. Although the Laplace equation (3.27a) for $\phi(x, z, t)$ is linear, the boundary conditions (3.27b) and (3.27c) required on the free surface contains nonlinear terms that cannot be written with the first power of the wave quantities η and ϕ. In addition, the fact that the position where these boundary conditions are imposed is the surface $z = \eta(x, t)$ deformed by the wave is another major factor that makes the problem nonlinear.

As mentioned in the derivation of the linearized system of basic equations (3.28), for example, $\phi(x, z, t)$ at $z = \eta(x, t)$ contained in (3.27c) is expressed by Taylor expansion around the undisturbed flat surface $z = 0$ as

$$\phi(x, z = \eta, t) = \phi(x, 0, t) + \eta \frac{\partial \phi(x, 0, t)}{\partial z} + \frac{1}{2!} \eta^2 \frac{\partial^2 \phi(x, 0, t)}{\partial z^2} + \cdots . \tag{3.75}$$

From this, the surface wave problem in which the boundary conditions are imposed on the surface deformed by the wave is a problem involving nonlinear terms of infinite order with respect to wave quantities such as η and ϕ.

The sinusoidal wave solution discussed so far is a solution of the linearized system (3.28) not of the original system (3.27) itself. Therefore, the sinusoidal wave solution is only an approx-imate solution. In order to use an approximate solution properly, it is important in general to know the magnitude of the error that the solution contains. Let us consider the kinetic boundary condition (3.27c) as an example. The process of linearization leaves η_t and ϕ_z in the equation

---

[13]To be honest, the author of this book himself is one such person.

but ignores the term $\phi_x \eta_x$. Also, even for the two terms left, they are replaced by the values at $z = 0$ from the values evaluated at $z = \eta$, and all the second and subsequent terms of the Taylor expansion (3.75) have been omitted. Evaluating the terms left using the linear sinusoidal wave solution

$$\eta_t(x, t) = \phi_z(x, z = 0, t) = -i\omega a\, e^{i(kx-\omega t)}. \tag{3.76}$$

Since we consider that the wave amplitude is small, the largest nonlinear term neglected by the linearization is the nonlinear term that is proportional to the square of the wave amplitude. Taking $\phi_x \eta_x$ as a representative of these, and similarly evaluating its magnitude by using the linear sinusoidal wave solution, it becomes

$$\phi_x(x, z = 0, t)\eta_x(x, t) = \frac{-ik\omega a^2}{\tanh kh}\, e^{2i(kx-\omega t)}. \tag{3.77}$$

Hence, the ratio of its magnitude to that of the linear term is given by

$$\frac{|\phi_x(x, z = 0, t)\eta_x(x, t)|}{|\eta_t(x, t)|} = \frac{ak}{\tanh kh}. \tag{3.78}$$

Similar results are obtained with other nonlinear terms that are proportional to the square of the amplitude. From this, the sinusoidal wave solution obtained as a result of linearization has validity only in the situation where

$$\frac{ak}{\tanh kh} \ll 1 \tag{3.79}$$

is satisfied.

For deep water waves where $kh \gg 1$, $\tanh kh \approx 1$ and (3.79) simply requires

$$ak = \frac{2\pi a}{\lambda} \ll 1, \tag{3.80}$$

implying that the amplitude is much smaller than the wavelength, or the slope of the surface waveform is much smaller than 1.

On the other hand, for shallow water waves where $kh \ll 1$, $\tanh kh \approx kh$, and the condition (3.79) becomes

$$ak \ll kh \ll 1. \tag{3.81}$$

This means that in order for the linear approximation to be allowable in shallow water waves, first the water depth h must be very small compared to the wavelength λ to be a shallow water wave, and in addition to that, the amplitude a must be very small compared to this small water depth h. From this, compared to the deep water wave for which the linear approximation is allowable if the amplitude is small compared to the wavelength, the situation where the linear approximation is valid for shallow water wave is much more limited. This suggests that nonlinear effects are more likely to appear for waves in shallow water.

Incidentally, the argument about the magnitude of a certain physical quantity becomes meaningful only after the reference material for comparison is specified. Until now, we have often used the ambiguous expression "assuming that the amplitude is small …" in the context of linear approximation, but the above argument teaches us its exact meaning.

By the way, the waves of the open ocean generated by the wind cannot grow so large because they break spontaneously and lose energy if they become too large. In real ocean waves, innumerable trains of waves with different wavelengths and directions of propagation are mixed, so it is not possible to apply the argument based on the sinusoidal wave solution as above, but according to the observation results, a statistically defined average amplitude corresponding to ak seems to be less than 0.1 in most cases. In this sense, it can be said that the ocean wave field is an object for which linear approximation is fairly effective.

3.4.2   STOKES WAVE: NONLINEAR STEADY TRAVELING WAVETRAINS

The above discussion has taught us the conditions for the linear approximation to hold, but at the same time, it gives us some hints on how to get a more accurate solution, including nonlinear effects, starting from a linear sinusoidal wave solution. Assuming for simplicity that the water depth h is infinite and the effect of surface tension can be ignored, let us consider how to find an approximate solution taking into account of nonlinearity.

As (3.77) illustrates, if the linear term is proportional to $a\,e^{i(kx-\omega t)}$, then the second-order nonlinear term has a form proportional to $a^2\,e^{2i(kx-\omega t)}$. From this, we can think of a method in which we assume an expansion form for η such as

$$\eta(x,t) = a\,e^{i\Phi} + \alpha_2 a^2\,e^{2i\Phi} + \alpha_3 a^3\,e^{3i\Phi} + \cdots, \quad \Phi = k(x - ct), \tag{3.82}$$

from the beginning, and determine the coefficients $\alpha_2, \alpha_3 \cdots$ such that (3.27) is satisfied. The suitable expansion for the velocity potential $\phi(x, z, t)$ corresponding to this form of η would be

$$\phi(x,z,t) = -\frac{i\omega_0 a}{k}\,e^{kz}e^{i\Phi} + \beta_2 a^2\,e^{2kz}e^{2i\Phi} + \beta_3 a^3\,e^{3kz}e^{3i\Phi} + \cdots, \tag{3.83}$$

where $\omega_0 = \sqrt{gk}$. Since the propagation velocity c may also deviate from the linear value c_0 due to nonlinear effects, we assume an expansion form in a also for c,

$$c = c_0 + c_1 a + c_2 a^2 + \cdots, \quad c_0 = \frac{\omega_0}{k} = \sqrt{\frac{g}{k}}. \tag{3.84}$$

The dependency e^{mkz} ($m = 1, 2, \ldots$) of ϕ on z is a consequence of the conditions that the ϕ with x dependence e^{imkx} must satisfy the Laplace equation (3.28a), and that ϕ must not diverge as $z \to -\infty$.

Substituting these expansions into the system of basic equations (3.27) and solving successively from the lower order of a, we obtain an approximate solution as follows [6]:

$$\eta = A\left[\cos\Phi + \frac{1}{2}\epsilon\cos 2\Phi + \frac{3}{8}\epsilon^2\cos 3\Phi + O(\epsilon^3)\right],$$

$$\phi = A\sqrt{\frac{g}{k}}\left[e^{kz}\sin\Phi + O(\epsilon^3)\right], \qquad c = \sqrt{\frac{g}{k}}\left[1 + \frac{1}{2}\epsilon^2 + O(\epsilon^3)\right], \qquad (3.85)$$

where $A = 2|a|$. $\epsilon \equiv Ak$ is a dimensionless parameter that indicates the smallness of nonlinearity, and $\epsilon \to 0$ corresponds to the linear approximation. An analysis for finding a steady traveling periodic wave solution including nonlinear effects by assuming a series expansion form consisting of harmonics of the linear sinusoidal wave solution as above was carried out for the first time by G. G. Stokes (1847). Therefore, an approximate solution like (3.85) is called the **Stokes wave**.[14]

The surface waveform of the Stokes wave corresponding to $\epsilon = 0.3$ is shown in Fig. 3.10. In the figure, "1st," indicates the waveform when only the fundamental wave (i.e., the linear sinusoidal wave), while "2nd" and "3rd" show the waveform when up to the 2nd and the 3rd harmonics are included, respectively. The Stokes wave solution (3.85) shows that the effects of nonlinearity on surface gravity waves appear in two aspects.

1. In the case of the linear sinusoidal waves, the surface displacement η is vertically symmetric with respect to the mean surface $z = 0$, but considering the nonlinear effect, the trough of the wave becomes shallower and flatter, and the crest becomes higher and more pointed as the figure shows, thus the waveform becomes vertically asymmetric.

2. For linear sinusoidal waves, the wave velocity is given by $c_0 = \sqrt{g/k}$ and depends only on the wavenumber (and hence the wavelength). However, considering nonlinear effects, a correction term that is proportional to $(Ak)^2$ appears in the wave velocity, and as a result, the wave with the same wavelength but with a larger amplitude has a slightly faster propagation velocity.[15]

The solution given by the Stokes expansion suggests that the crest becomes sharper and sharper as the amplitude increases, and the crest may become angular in the limiting condition. Stokes showed that if the crest has an angle in such a limiting condition, the angle must be 120° in order to satisfy the boundary conditions at the free surface ((3.27b) with $\tau = 0$ and (3.27c)) theoretically using the complex function theory. (See for example Section 13.13 of [7].)

At present, the concrete waveform of this limiting Stokes wave with an angle of 120°, that is, the steady traveling periodic wave with the maximum amplitude, is obtained very accurately

---

[14]It is that Stokes that leaves his name in the "Navier–Stokes equation" of fluid mechanics, and also in the "Stokes theorem" of vector analysis.

[15]As will be dealt with in detail in Chapter 6, this amplitude dependence of the wave velocity causes the phenomenon of "modulational instability" of the wavetrain, which also leads to interesting (and sometimes dangerous) phenomena such as the generation of giant waves in the ocean.

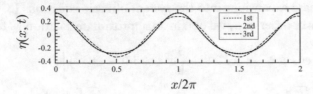

Figure 3.10: The surface profile of the Stokes wave ($k = 1.0$, $A = 0.3$).

by numerical calculations, and it is known that its amplitude is $Ak = 0.4434$ for infinite water depth, so that the ratio of wave height to wavelength is about 1/7.

For more detailed information on the overall contents of this chapter, refer to, for example, [1, 3, 5, 7] and Chapter 7 of [4].

COFFEE BREAK: LIGHT AND SOUND ARE NON-DISPERSIVE

When there is no dispersion, that is, the dispersion relation is $\omega = c_0 k$ (c_0 is a constant), waves of all wavelengths and frequencies travel at the same speed c_0. Therefore, given an initial waveform, the various wavelength components that compose it are all transmitted at the same speed, so the initial waveform is transmitted at a wave velocity c_0 without distortion. Light and sound are examples of such non-dispersive waves.

When asked "What is the speed of light?" many people will immediately answer "300,000 km/s" or "7 and a half laps of the earth in one second." At this time, the condition of "what color of light ⋯" is not attached. "Light" is a term that refers to the electromagnetic waves visible to humans, and its wavelength is approximately $(4 \sim 8) \times 10^{-7}$ m. Of these, the short end of the wavelength is purple and the long end is red, and ultraviolet shorter than violet and infrared longer than red are invisible to humans. Some members of electromagnetic waves have much longer and shorter wavelengths than visible light. For example, the wavelength of an electromagnetic wave called "microwave" used for heating things in a microwave oven is about 12 cm, the wavelength of the electromagnetic waves transmitted from radio broadcasting stations for AM (Amplitude Modulation) in Japan are several hundred meters, which is far longer than visible light. On the other hand, X-rays used in hospital radiography is also a kind of electromagnetic waves, but their wavelength is less than 10^{-9} m = 1 nm, less than 1/100 of visible light, and compared to the radio waves of AM radio, it is less than 1/100 billion. Nevertheless, electromagnetic waves are non-dispersive waves, and therefore, regardless of the wavelength, whether they are X-rays or radio waves from broadcasting stations, the propagation velocity is the same at about 300,000 km/s.[16]

---

[16]However, the fact that electromagnetic waves, including visible light, are non-dispersive is limited to being in vacuum. In materials, they become dispersive and differences in propagation speed appear depending on the wavelength. The fact that the light can be divided into the seven colors of the rainbow by a glass prism is exactly that manifestation. The speed of light

Sound wave is also one of the most familiar wave phenomena to us. As in the case of the speed of light, if you are asked about the speed of sound, many of you will reply, "about 340 m/s." If you remember a little more well, you may answer "331 + 0.6× temperature (°C)." Even at this time, nobody would ask back, "What is the frequency of the sound?" This is also because sound waves are non-dispersive waves.

The audible range of frequency that can be heard by the human ear is said to be approximately 20–20,000 Hz. Within this range alone, there is a difference of 1000 times. In addition to the audible range, there are ultrasonic waves that humans cannot hear but bats can, and very low frequency sounds that are not directly audible to humans but known to have some bad effect on human's health. If you include these, sound waves have a fairly wide range of wavelengths. However, all these frequency components propagate at the same speed, so the initial waveform does not change with propagation. The reason why we can use "voice," that is, sound wave as a tool of daily communication is that sound wave is non-dispersive. If the sound wave were dispersive, what would happen? When you say to your friend "I love you," the various frequency components included in the pressure fluctuation generated by your voice would be transmitted at different speeds, the waveform would change as it propagates, and it would become just a noise without meaning when your voice arrives at the ear of your friend. Suppose that the sound wave has the same dispersive character as the water wave, that is, the lower the frequency, the faster it travels. Then, even if all the instruments of the orchestra are playing at the same time in a concert hall, the sound of the low sound of contrabass would reach first to the audience and the high sound of piccolo would arrive later. Then a great performance would turn into nothing but a mess.

In order for waves to be non-dispersive, the only permitted form of the dispersion relation is $\omega(k) = c_0 k$. In this sense, non-dispersive waves are quite special waves. Bearing this in mind, we should be more grateful to the "accidental fortune" that the sound wave, i.e., the wave of density and pressure which is completely governed by the laws of physics, is not dispersive.

3.5   REFERENCES

[1] G. D. Crapper. *Introduction to Water Waves*. Ellis Horwood, 1984. 62

[2] H. Goldstein. *Classical Mechanics*. Addison-Wesley, 1980. DOI: 10.2307/3610571 55

[3] R. S. Johnson. *A Modern Introduction to the Mathematical Theory of Water Waves*. Cambridge U.P., 1997. DOI: 10.1017/cbo9780511624056 62

[4] P. K. Kundu and I. M. Cohen. *Fluid Mechanics*, 3rd ed., Elsevier, 2004. 62

in vacuum divided by the speed of light in a material is called the **refractive index** of the material. The larger the refractive index of a material, the more the light is bent toward the material when the light enters it. In glass, violet light with a shorter wavelength propagates at a slower speed than red light with a longer wavelength, so it is more refracted when the light is incident on the glass.

[5] J. Lighthill. *Waves in Fluids*. Cambridge U.P., 1978. DOI: 10.1002/cpa.3160200204 55, 58, 62

[6] J. V. Wehausen and E. V. Laitone. Surface wave. In S. Flügge, Ed., *Encyclopedia of Physics*, vol. 6, Springer, 1960. DOI: 10.1007/978-3-642-45944-3_6 61

[7] G. B. Whitham.   *Linear and Nonlinear Waves*.   John Wiley & Sons, 1974. DOI: 10.1002/9781118032954 61, 62

CHAPTER 4

Perturbation Method and Multiple Scale Analysis

Perturbation method is a method for finding approximate solutions in the form of a power series in ϵ when a small parameter ϵ is included in the equation to be solved. The basic system of equations for water surface waves (3.27) does not include a small parameter explicitly. But, for example, in the situation of long wave in which the wavelength λ is much longer than the water depth h (i.e., $h/\lambda \ll 1$) or in the situation of quasi-monochromatic wavetrain to be dealt with in Chapter 6 in which the width of the spectrum Δk is much smaller than the carrier wavenumber k_c (i.e., $\Delta k/k_c \ll 1$), small parameters appear when the equations are non-dimensionalized in an appropriate manner. By using the perturbation method that makes use of the smallness of the parameter, it is possible to obtain an approximate solution or to reduce the equation itself to a simpler one. Thus, the perturbation method will play an important role in many parts of this book in the future, so the basic knowledge about it will be summarized in this chapter. We also introduce the "multiple scale method," which is an evolved perturbation method that remains effective in the situation where simple perturbation methods fail.

4.1 NECESSITY OF APPROXIMATE SOLUTION METHOD

Consider again a simple pendulum, as shown in Fig. 3.9. Newton's law of motion for tangential direction can be written as

$$ml\frac{d^2\theta}{dt^2} = -mg\sin\theta. \tag{4.1}$$

This equation is a nonlinear ordinary differential equation and is difficult to solve.[1] What makes this equation nonlinear is $\sin\theta$ on the right-hand side. $\sin\theta$ contains a power of infinite order of θ, as shown by its Taylor expansion

$$\sin\theta = \theta - \frac{\theta^3}{3!} + \frac{\theta^5}{5!} - \cdots, \tag{4.2}$$

---

[1]In the case of this equation, it can be solved analytically if you have knowledge about the "elliptic functions." In that case, the exact solution is given by $\theta(t) = 2\sin^{-1}\left[k\,\text{sn}(\omega_0 t, k)\right], k = \sin(\theta_{\max}/2)$, where $\text{sn}(x, k)$ is one of Jacobi's elliptic functions.

and is not a linear function.

However, if the swing angle is small enough to approximate $\sin \theta$ by θ, the equation to be solved is a linear differential equation

$$ml\frac{d^2\theta}{dt^2} = -mg\,\theta \longrightarrow \frac{d^2\theta}{dt^2} = -\omega_0^2\,\theta, \quad \omega_0 \equiv \sqrt{g/l}, \tag{4.3}$$

and the general solution can be easily obtained as $\theta(t) = A\cos(\omega_0 t + \phi)$. Here, A and ϕ are arbitrary constants, A is the amplitude, and ϕ is the initial phase. The solution shows that the frequency ω_0 of the pendulum with a small swing angle is determined by the ratio of the length l to the gravitational acceleration g by $\omega_0 = \sqrt{g/l}$, and it does not depend on the mass m or the amplitude A. This property is known as "isochronism" of a pendulum. Thus, assuming that the amplitude is small and approximating it with a linear equation, the solution becomes much easier. However, it is known that in a pendulum governed by the original equation (4.1), isochronism does not hold, and the period becomes longer as the amplitude becomes larger. That is, the important property that the period depends on the amplitude has been lost by the linearization.

Not only for the pendulum, but also for various phenomena around us, the rules governing them are often expressed in the form of nonlinear differential equations. However, most of these nonlinear differential equations cannot be solved analytically. The pendulum, which can be solved by using the sophisticated elliptic functions, is rather exceptional. If there is no means to obtain exact solutions in most cases, it is important to find approximate solutions that reproduce, albeit incomplete, the important parts of the properties possessed by the true solutions. One of the representative methods for obtaining such approximate solutions is the perturbation method.

4.2   PERTURBATION METHOD

Perturbation method is a method for finding approximate solutions in the form of a series expansion in ϵ when a small parameter ϵ is included in the equation to be solved.[2]

4.2.1   APPROXIMATE VALUE OF ROOT OF QUADRATIC EQUATION

Let's try one exercise first. For a quadratic equation

$$x^2 + \epsilon x - 1 = 0, \tag{4.4}$$

---

[2]"Perturbation" is a term derived from astronomy. For example, when considering the movement of the earth around the sun, Newton's equation of motion considering only the attraction between the sun and the earth can be solved exactly (two-body problem), and the earth draws an elliptical orbit with the sun as one focus. However, in reality, although it is weaker than the gravitational force from the sun, there is also an attraction from other planets such as Mars and Jupiter, which disturbs the elliptical orbit of the earth. The disturbance of the orbit due to the minute influence from celestial bodies other than the two objects in question has long been called as "perturbation" in astronomy.

consider $\epsilon \ll 1$ and find an approximation of the root. Of course, if we use the formula for the roots of the quadratic equation, exact roots can be easily obtained as

$$x = \frac{-\epsilon \pm \sqrt{\epsilon^2 + 4}}{2}, \tag{4.5}$$

but in order to show the procedure of the perturbation method, we try to find an approximate value of the roots by perturbation method.

First, consider approximating the root x with a series in the small parameter ϵ as

$$x(\epsilon) = x_0 + \epsilon x_1 + \epsilon^2 x_2 + \cdots. \tag{4.6}$$

Since $x = x_0$ when $\epsilon = 0$, x_0 must be the root for $\epsilon = 0$, that is,

$$x_0^2 - 1 = 0, \tag{4.7}$$

so $x_0 = \pm 1$. Let us choose $x_0 = 1$ of these and find out how this is modified when $\epsilon \neq 0$. Assuming

$$x(\epsilon) = 1 + \epsilon x_1 + \epsilon^2 x_2 + \cdots, \tag{4.8}$$

and inserting to (4.4) gives

$$(1 + \epsilon x_1 + \epsilon^2 x_2 + \cdots)^2 + \epsilon(1 + \epsilon x_1 + \epsilon^2 x_2 + \cdots) - 1 = 0. \tag{4.9}$$

Arranging this equation according to the power of ϵ gives

$$(1 - 1) + (2x_1 + 1)\epsilon + (2x_2 + x_1^2 + x_1)\epsilon^2 + O(\epsilon^3) = 0. \tag{4.10}$$

Since ϵ is a parameter that can take any value as long as it is small, the coefficients of each order in ϵ on the left side must be all 0 in order for this equation to hold. Then, $x_1 = -\frac{1}{2}$ from the coefficient of ϵ^1, substituting this into the coefficient of ϵ^2 and requiring it to vanish, we obtain $x_2 = \frac{1}{8}$. Substituting these values of x_1 and x_2 into the coefficient of ϵ^3, then we obtain $x_3 = \cdots$. In this way, the coefficients of the expansion (4.8) are determined successively from the lower order by a simple calculation. As a result, we obtain

$$x = 1 - \frac{1}{2}\epsilon + \frac{1}{8}\epsilon^2 + O(\epsilon^3), \tag{4.11}$$

as an approximation to the root close to 1 among two roots of (4.4).

When $\epsilon = 0.1$, for example, the exact value given by the quadratic formula is $0.9512492\cdots$, while the approximate solution (4.11) gives 0.95125, with an error of less than a millionth. Even when $\epsilon = 1.0$, which seems ridiculously large though, the approximate solution (4.11) gives 0.625 for the exact root $0.6180339\cdots$, and the relative error is only about 1%. When using the perturbation method, we formally require $\epsilon \ll 1$, but, as this example shows, there are many cases where the obtained results remain unexpectedly accurate for surprisingly large ϵ.

4.2.2   APPROXIMATE SOLUTION OF DIFFERENTIAL EQUATION

We can also solve differential equations approximately using perturbation methods. For an exercise, let us find the approximate solution of the initial value problem of a nonlinear ordinary differential equation

$$\dot{x} + x + \epsilon x^2 = 0, \qquad x(0) = 1 \tag{4.12}$$

by the perturbation method.

As in the above example, we start with a series for x

$$x(t) = x_0(t) + \epsilon x_1(t) + \epsilon^2 x_2(t) + \cdots . \tag{4.13}$$

What differs from the above example is that the expansion coefficient x_i is not a constant but a function of t. From the initial condition

$$x(0) = x_0(0) + \epsilon x_1(0) + \epsilon^2 x_2(0) + \cdots = 1, \tag{4.14}$$

we obtain the initial conditions for each $x_i(t)$,

$$x_0(0) = 1, \quad x_1(0) = 0, \quad x_2(0) = 0, \cdots . \tag{4.15}$$

Substituting (4.13) into (4.12) gives

$$(\dot{x}_0 + \epsilon \dot{x}_1 + \epsilon^2 \dot{x}_2 + \cdots) + (x_0 + \epsilon x_1 + \epsilon^2 x_2 + \cdots) + \epsilon(x_0 + \epsilon x_1 + \epsilon^2 x_2 + \cdots)^2 = 0. \tag{4.16}$$

If we arrange this by powers of ϵ and solve it from the lower order,

$$O(\epsilon^0): \quad \dot{x}_0 + x_0 = 0, \quad x_0(0) = 1 \longrightarrow x_0(t) = e^{-t}, \tag{4.17a}$$

$$O(\epsilon^1): \quad \dot{x}_1 + x_1 = -x_0^2 = -e^{-2t}, \quad x_1(0) = 0 \longrightarrow x_1(t) = -e^{-t} + e^{-2t}, \tag{4.17b}$$

$$O(\epsilon^2): \quad \dot{x}_2 + x_2 = -2x_0 x_1 = 2e^{-2t} - 2e^{-3t}, \quad x_2(0) = 0,$$

$$\longrightarrow \quad x_2(t) = e^{-t} - 2e^{-2t} + e^{-3t}, \tag{4.17c}$$

and we get

$$x(t) = e^{-t} + \epsilon \left(-e^{-t} + e^{-2t}\right) + \epsilon^2 \left(e^{-t} - 2e^{-2t} + e^{-3t}\right) + O(\epsilon^3) \tag{4.18}$$

as an approximate solution up to $O(\epsilon^2)$.

In fact, (4.12) is a nonlinear differential equation called "Bernoulli type," which can be reduced to a linear differential equation by a simple variable transformation. Moreover, since it is also a separable differential equation, it is possible to solve it by thinking so. The exact solution is given by

$$x(t) = \frac{1}{(1 + \epsilon)e^t - \epsilon}. \tag{4.19}$$

Although the error included in the approximate solution (4.18) gradually increases with time t, it can be confirmed that, for example, when $\epsilon = 0.1$, the relative error remains less than 0.1% at $t = 10$. Also, if we consider the exact solution (4.19) as a function of ϵ and find the Maclaurin expansion (i.e., Taylor expansion around $\epsilon = 0$) in ϵ, it agrees with the approximate solution (4.18) obtained by the perturbation method.

4.3   APPLICATION TO NONLINEAR PENDULUM

4.3.1   BREAKDOWN OF REGULAR PERTURBATION METHOD

From the above examples, it was found that even for nonlinear problems that cannot be solved exactly, it may be possible to obtain effective approximate solutions with relatively simple calculations by using the perturbation method. But, unfortunately, it does not always work this way. In fact, as we will see below, even with the simple problem of a single pendulum mentioned at the beginning of this chapter, the usual perturbation method cannot obtain an effective approximate solution.

Since a small parameter does not appear explicitly in the pendulum equation (4.1), it is first necessary to rewrite the equation so that the perturbation method can be used, that is, a small parameter appears explicitly. Let the maximum swing angle of the pendulum be $\theta_{\max}$ and introduce a dimensionless variable x by $x = \theta/\theta_{\max}$. Also, let us introduce a dimensionless time variable $\tilde{t}$ by $\tilde{t} = \omega_0 t$, and express the derivative of x with respect to $\tilde{t}$ as $\dot{x}$. Substituting the Maclaurin series (4.2) of $\sin \theta$ to (4.1) and introducing ϵ by $\epsilon = \theta_{\max}^2/6$, (4.1) becomes

$$\ddot{x} + x - \epsilon x^3 + \frac{3}{10}\epsilon^2 x^5 - \cdots = 0, \quad x(0) = 1, \quad \dot{x}(0) = 0. \tag{4.20}$$

Here, in order to be specific, it is assumed that the pendulum is pulled up to $\theta_{\max}$ at $t = 0$ and released with speed 0. In the linear approximation that predicts the pendulum isochronism, all terms that contain ϵ on the left side are ignored. However, in order to obtain an approximate solution including nonlinear effects, we treat ϵ as a small but nonzero parameter, and try to find an approximate solution by the perturbation method.

First of all, assume as usual

$$x(t) = x_0(t) + \epsilon x_1(t) + \epsilon^2 x_2(t) + \cdots. \tag{4.21}$$

Since there is no fear of confusion, we will omit the tilde from $\tilde{t}$ and simply write t. From the initial condition,

$$x(0) = x_0(0) + \epsilon x_1(0) + \cdots = 1, \quad \dot{x}(0) = \dot{x}_0(0) + \epsilon \dot{x}_1(0) + \cdots = 0. \tag{4.22}$$

This gives the initial condition for each $x_i(t)$ as follows:

$$x_0(0) = 1, \quad x_i(0) = 0 \ (i = 1, 2, \cdots), \quad \dot{x}_i(0) = 0 \ (i = 0, 1, 2, \cdots). \tag{4.23}$$

Substituting (4.21) into (4.20),

$$\left(\ddot{x}_0 + \epsilon\ddot{x}_1 + \epsilon^2\ddot{x}_2 + \cdots\right) + \left(x_0 + \epsilon x_1 + \epsilon^2 x_2 + \cdots\right)$$
$$- \epsilon\left(x_0 + \epsilon x_1 + \epsilon^2 x_2 + \cdots\right)^3 + \frac{3}{10}\epsilon^2\left(x_0 + \epsilon x_1 + \epsilon^2 x_2 + \cdots\right)^5 - \cdots = 0. \qquad (4.24)$$

Arranging this according to the power of ϵ and solving successively from the lower order, we obtain

$$O(\epsilon^0): \quad \ddot{x}_0 + x_0 = 0, \quad x_0(0) = 1, \quad \dot{x}_0(0) = 0 \quad \longrightarrow \quad x_0(t) = \cos t, \qquad (4.25a)$$

$$O(\epsilon^1): \quad \ddot{x}_1 + x_1 = x_0^3 = \cos^3 t = \frac{1}{4}\cos 3t + \frac{3}{4}\cos t, \quad x_1(0) = 0, \quad \dot{x}_1(0) = 0$$

$$\longrightarrow \quad x_1(t) = \frac{1}{32}\cos t - \frac{1}{32}\cos 3t + \frac{3}{8}t\sin t. \qquad (4.25b)$$

Therefore, we obtain

$$\tilde{x}(t) = \cos t + \epsilon\left(\frac{1}{32}\cos t - \frac{1}{32}\cos 3t + \frac{3}{8}t\sin t\right) + O(\epsilon^2) \qquad (4.26)$$

as an approximate solution $\tilde{x}(t)$ considering up to $\epsilon x_1(t)$. However, the term $\frac{3}{8}t\sin t$ that appears in $O(\epsilon)$ causes a trouble here.

When we assume that the solution is represented by an infinite series like (4.21), it is implicitly assumed that we are to obtain only the first few terms and then truncate the series, and we have no idea of obtaining until the end of the infinite series. In order to obtain a reasonable approximate solution no matter where we truncate the series, the relation $\epsilon^n x_n \gg \epsilon^{n+1} x_{n+1}$ must hold for any n, that is, the first term ignored must be much smaller than the last term left.[3] The series (4.21) satisfies this property under the condition of $\epsilon \ll 1$ if the magnitude of each coefficient $x_n(t)$ remains of $O(1)$. Looking at the approximate solution (4.26), the part $x_0(t) = \cos t$ always remains $O(1)$. In the part multiplied by ϵ, the first and the second term always remain $O(\epsilon)$ and do not cause any problem, but the third term $\epsilon\frac{3}{8}t\sin t$ grows infinitely with time, no matter how small ϵ is. A term that grows infinitely with time in this way is called a **secular term**.

When time t becomes about $1/\epsilon$, the first term x_0 and the second term $\epsilon x_1(t)$ of the series (4.21) become similar in magnitude, and the rationality of the approximation is lost. For example, let $\epsilon = 0.01$. This corresponds to $\theta_{max} \approx 0.25$ (rad) about 14°) in amplitude. The assumption that $\epsilon \ll 1$ seems to be not so bad for this value of ϵ, but even in this case, if t is about $1/\epsilon = 100$, the approximate solution (4.26) becomes completely useless. According to the linear theory, the period of this pendulum is 2π, so $t = 100$ is only 16 times of swing of the pendulum. Figure 4.1 shows a comparison of the approximate solution (4.26) and the exact solution when $\theta_{max} = 45°$ ($\epsilon \approx 0.1$). It can be observed that the exact solution (solid line) oscillates at a constant amplitude, while the approximate solution (4.26) obtained by the perturbation method (dashed line) becomes larger and larger with time due to the secular term.

---

[3]A series with this property is called an **asymptotic series**.

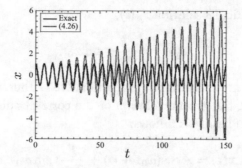

Figure 4.1: Behavior of the approximate solution (4.26). Dotted line: the approximate solution (4.26); solid line: exact solution. ($\theta_{\max} = 45°$, $\epsilon \approx 0.103$.)

4.3.2 FORCED OSCILLATION AND RESONANCE

Why does the perturbation method, which worked so well for the approximation of roots of quadratic equation and the calculation of approximate solution of the first-order differential equation, not work well for the nonlinear pendulum problem? The direct cause is that, as we saw above, the solution of $O(\epsilon)$ has an infinitely growing secular term. But in the background of the appearance of this secular term is a phenomenon called **resonance**.

Let us consider the equation of forced oscillation of a linear oscillator driven by a sinusoidal external force,

$$\ddot{x} + \omega_0^2 x = F \cos \omega t, \tag{4.27}$$

where ω_0 is the natural frequency of the oscillator, and ω, F are the frequency and the amplitude of the external force, respectively. If $\omega \neq \omega_0$, i.e., if the frequency of the external force is different from the natural frequency, the general solution to this equation is given by

$$x(t) = A \cos(\omega_0 t + \phi) + \frac{F}{\omega_0^2 - \omega^2} \cos \omega t, \tag{4.28}$$

where A, ϕ are certain constants determined from the initial conditions. (4.28) shows that the oscillation realized in this situation is a superposition of harmonic oscillation with the natural frequency ω_0 of the oscillator itself and the harmonic oscillation with the frequency ω of the external force. It also shows that the amplitude of the part oscillating with the external frequency is proportional to the strength F of the external force and increases as ω^2 and ω_0^2 become closer. The amplitudes of the two harmonic oscillations that make up the solution are constant, and $x(t)$ remains finite forever.

On the other hand, in the case of $\omega = \omega_0$, (4.27) becomes

$$\ddot{x} + \omega_0^2 x = F \cos \omega_0 t, \tag{4.29}$$

and this equation has a particular solution $x_p(t)$,

$$x_p(t) = \frac{F}{2\omega_0} t \sin \omega_0 t. \tag{4.30}$$

Note that the amplitude is not constant but proportional to t. Thus, the general solution in this case is given as the sum of the general solution of the corresponding homogeneous equation (i.e., no external forcing) and $x_p(t)$ as follows:

$$x(t) = A \cos(\omega_0 t + \phi) + \frac{F}{2\omega_0} t \sin \omega_0 t. \tag{4.31}$$

Thus, when an oscillator is driven by an oscillatory external force with the same frequency as its natural frequency, the oscillation amplitude increases without limit. This phenomenon is called **resonance**.

In June 2000, there was an event that the Millennium Bridge in London was shut down on the second day of its operation. The Millennium Bridge, as the name suggests, was a wonderful pedestrian-only bridge built with a huge expense of over 18 million pounds to commemorate the arrival of the 21st century, but when a large number of pedestrians walked it caused a big shake, and it was temporarily closed for repair. This repair work costed extra 5 million pounds and the bridge was opened again in February 2002. Since humans walk on two legs, the center of gravity of the body swings slightly to the left and right when taking a step forward. At this time, humans exert lateral force on the bridge. The force that one person exerts on a huge bridge is negligible. However, if the bridge starts to shake a little for some reason (crosswind, for example), many people on it (some 2,000 people at one time) will tend to keep pace with the shake and take a step because it is easier to walk if it is adjusted to the bridge's shaking. When this happens, the frequency of the bridge's roll and the lateral force applied to the bridge by many people will match, a resonance phenomenon will occur, and the shake will get bigger and bigger. It is now thought that this resonance phenomenon is the cause of the vibration problem of the Millennium Bridge.

In the process of applying the perturbation method to the nonlinear pendulum problem to obtain an approximate solution, this resonance actually occurred. That is, if we look at (4.25) for $x_1(t)$, it can be seen that the oscillator with natural frequency $\omega_0 = 1$ is forced by the oscillatory external force with the same frequency. Because of this, a resonance occurs and a secular term $t \sin t$ appears, and the approximate solution breaks down.

4.4   MULTIPLE SCALE ANALYSIS

4.4.1   MULTIPLE TIME SCALE

So, is the perturbation method totally useless even with this simple problem of nonlinear pendulum? Isn't there any good idea to make the perturbation method useful again? Thanks to the efforts of previous researchers, a group of "evolved" perturbation methods has been developed that do not lose their effectiveness even when the regular perturbation method breaks down. They are called the **singular perturbation methods**. Here, we take up the **multiple scale method**, which is one of the most easily accessible and versatile among them, and introduce its concept and effectiveness by applying it to the problem of nonlinear pendulum.

Let us consider again the nonlinear pendulum equation (4.20). The first two terms without ϵ are far larger, and the basic part of the motion is determined by them. This is the harmonic oscillation of $\omega_0 = 1$ given by the linear theory. Comparing with them, the terms with ϵ after the third term are much smaller. In order to describe the effects of these minor factors (for example, deviation from isochronism), it will be necessary to observe the motion for a fairly long time.

That is, there exits two important time scales in the motion of the nonlinear pendulum. One is the usual time which treats each swing of the pendulum. The t used in the equation (4.20) is a time variable that is well suited to observing this each swing of the pendulum. The other is a long time when nonlinear effects become visible. Since the magnitude of the terms that produce nonlinear effects is $O(\epsilon)$, it may take as long as $1/\epsilon$ for them to accumulate and become visible. This also agrees with the fact that the time t at which the approximate solution obtained by the regular perturbation method breaks down due to the growth of the secular term is about $1/\epsilon$.

For example, when talking about 100 m runs of athletics, we use seconds as units, but when talking about global climate change, we use units of 100 years or 1000 years, even though there is one and the same flow of time. Similarly, it is not surprising that the idea of introducing two time variables at the same time has arisen if the phenomenon to be treated contains two very different important times. This is the idea called the **multiple time scale**.

In the multiple time scale method, a new time variable τ is introduced by $\tau = \epsilon t$, in addition to the time variable t initially present. Since this new time variable τ only changes by about 1 after a long time of about $1/\epsilon$ when viewed with the original time variable t, it is a suitable time variable to see a long time when nonlinear effects appear. And we treat the dependent variable $x(t)$, which is originally a function of only t, as a function of two independent time variables like $x(t, \tau)$. Of course, because $\tau = \epsilon t$, t and τ are not independent, and treating x like this as a two-variable function is just a kind of trick. However, as we will see below, this little trick can avoid the troubles faced by the regular perturbation methods. That is, by taking advantage of the new "degree of freedom" created by considering x, which is a function of only t, as if it were a function of t and τ, it is possible to avoid the occurrence of the secular term that makes the regular perturbation method break down. In the following, we will show the procedure of multiple scale method specifically for (4.20).

4.4.2   APPLICATION OF MULTIPLE TIME SCALE TO NONLINEAR PENDULUM

Let us introduce a new time variable τ by $\tau = \epsilon t$. At the same time, to distinguish t in multiple time scale analysis from the original single time t, we write t as t_0. Then $x(t)$ is treated as $x(t_0, \tau)$.

Then, the original time derivative changes to

$$\frac{dx}{dt} = \frac{\partial x}{\partial t_0}\frac{dt_0}{dt} + \frac{\partial x}{\partial \tau}\frac{d\tau}{dt} = \frac{\partial x}{\partial t_0} + \epsilon\frac{\partial x}{\partial \tau} \tag{4.32}$$

according to the chain rule of partial derivatives. That is, if only the part of the differential operation is extracted, the following replacement occurs:

$$\frac{d}{dt} = \frac{\partial}{\partial t_0} + \epsilon\frac{\partial}{\partial \tau}. \tag{4.33}$$

Substituting this into (4.20) and representing x as a series expansion

$$x(t) = x(t_0, \tau) = x_0(t_0, \tau) + \epsilon x_1(t_0, \tau) + \cdots , \tag{4.34}$$

as usual gives

$$\left(\frac{\partial}{\partial t_0} + \epsilon\frac{\partial}{\partial \tau}\right)^2 (x_0 + \epsilon x_1 + \cdots) + (x_0 + \epsilon x_1 + \cdots) - \epsilon (x_0 + \epsilon x_1 + \cdots)^3 + \cdots = 0. \tag{4.35}$$

We arrange this according to the power of ϵ and solve it from the lower order. At first for $O(\epsilon^0)$,

$$\frac{\partial^2 x_0}{\partial t_0^2} + x_0 = 0 \quad \longrightarrow \quad x_0(t_0, \tau) = A(\tau)\cos[t_0 + \phi(\tau)]. \tag{4.36}$$

If we were using the conventional regular perturbation method, the amplitude A and the phase constant ϕ were constants determined from the initial conditions. However, in the multiple time scale method, they are constant with respect to the fast time t_0, but are permitted to be functions of the slow time τ.

At the next order $O(\epsilon^1)$, we obtain

$$\frac{\partial^2 x_1}{\partial t_0^2} + x_1 = x_0^3 - 2\frac{\partial^2 x_0}{\partial t_0 \partial \tau}. \tag{4.37}$$

Substituting the solution of $O(\epsilon^0)$ into x_0 gives

$$\frac{\partial^2 x_1}{\partial t_0^2} + x_1 = A^3 \cos^3(t_0 + \phi) + 2\frac{\partial}{\partial \tau}[A\sin(t_0 + \phi)]$$

$$= \frac{1}{4}A^3 \cos[3(t_0 + \phi)] + \frac{3}{4}A^3 \cos(t_0 + \phi) + 2\frac{\partial A}{\partial \tau}\sin(t_0 + \phi)$$

$$+ 2A\cos(t_0 + \phi)\frac{\partial \phi}{\partial \tau}. \tag{4.38}$$

Comparing with the corresponding equation (4.25b) in the regular perturbation method, it can be seen that two terms including τ derivative are newly added to the right-hand side, as a result of A and ϕ being allowed to depend on the slow time τ. Of the right side of (4.38), the resonant "external force" that cause the generation of secular terms are the terms of $\sin(t_0 + \phi)$ and $\cos(t_0 + \phi)$. Therefore, the following conditions are obtained as conditions with which these terms disappear and no secular term occurs:

$$\sin(t_0 + \phi): \quad \frac{\partial A}{\partial \tau} = 0 \quad \longrightarrow \quad A(\tau) = A_0 \ (A_0 = \text{const.}), \tag{4.39}$$

$$\cos(t_0 + \phi): \quad \frac{\partial \phi}{\partial \tau} + \frac{3}{8}A^2 = 0 \quad \longrightarrow \quad \phi(\tau) = -\frac{3}{8}A_0^2\tau + \phi_0 \ (\phi_0 = \text{const.}). \tag{4.40}$$

Such a condition as above for preventing the occurrence of a secular term is called a **non-secular condition**. If the amplitude A and the phase constant ϕ change slowly to satisfy this condition, no secular term occurs, x_1 remains $O(1)$, and therefore the series approximation (4.34) does not break down. Then, from (4.36), the most dominant part of the solution x_0 is given by

$$x_0 = A(\tau)\cos[t_0 + \phi(\tau)] = A_0 \cos\left(t_0 - \frac{3}{8}A_0^2\,\tau + \phi_0\right) = A_0 \cos\left[\left(1 - \frac{3}{8}\epsilon A_0^2\right)t + \phi_0\right]. \tag{4.41}$$

In (4.41), if we change the time variable back to the original variable,[4] substitute the definition of ϵ, i.e., $\epsilon = \theta_{\max}^2/6$, and consider that $x = 0$ at $t = 0$, then we get

$$\theta(t) = \theta_{\max} \cos\left[\omega_0\left(1 - \frac{1}{16}\theta_{\max}^2\right)t\right] + O(\epsilon). \tag{4.42}$$

This approximate solution shows that the frequency ω and the period T of the pendulum are given by

$$\omega \approx \omega_0\left(1 - \frac{1}{16}\theta_{\max}^2\right), \qquad T = \frac{2\pi}{\omega} \approx T_0\left(1 - \frac{1}{16}\theta_{\max}^2\right)^{-1}, \tag{4.43}$$

implying that the frequency decreases and the period gets longer with the increase of the amplitude $\theta_{\max}$.

Figure 4.2 depicts the ratio of the period T to the linear period $T_0 (= 2\pi/\omega_0)$ as a function of the amplitude $\theta_{\max}$. The solid line shows the result based on the exact solution, and the dotted line shows the result (4.43) obtained by the multiple time scale method, i,e,, $T/T_0 = 1/\left(1 - \theta_{\max}^2/16\right)$. It can be seen that, even though it is an approximation that incorporates only minimal nonlinear effects, as $\theta_{\max}$ increases to as much as 150° it still gives quite a good prediction.

Figure 4.3 shows the comparison between the linear solution (dotted line) without considering nonlinear effects and the exact solution (solid line) when $\theta_{\max} = 45°$, the same case as that shown in Fig. 4.1. Since the linear solution cannot take into account the elongation of the period resulting from the nonlinear effect, the phase difference with the exact solution increases

---

[4]When (4.20) is derived, $\tilde{t}$ is introduced by $\tilde{t} = \omega_0 t$. Since we have omitted the tilde since then, t here is correctly $\tilde{t} = \omega_0 t$.

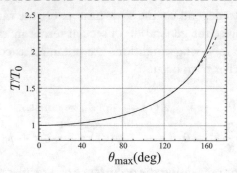

Figure 4.2: Period T of a single pendulum vs. amplitude θ_{max}. Solid line: exact result; dotted line: approximate result (4.43).

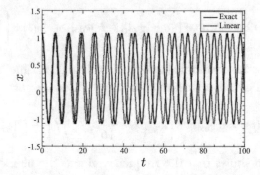

Figure 4.3: Exact solution (solid line) vs. linear solution (dashed line). ($\theta_{max} = 45°$.)

with time, and in just 10 oscillations or so, the phase difference grows so large that the linear solution predicts the pendulum to swing to the left although the exact solution predicts it swing to the right. On the other hand, Fig. 4.4 shows the comparison between the approximate solution (4.42) obtained by the multiple time scale method and the exact solution for the same case. Despite the relatively large amplitude of $\theta_{max} = 45°$, the approximate solution agrees with the exact solution so well that the difference can hardly be seen, at least as far as the time of about 16 oscillations shown in the figure.

Although it is a tricky idea of "introducing multiple time variables and treating them as if they were independent variables," this example clearly shows that this methodology is very simple and powerful. The multiple scale method is often used in this book from now on. For more details and applications of the perturbation methods in general, refer to [1, 2] and [3], for example.

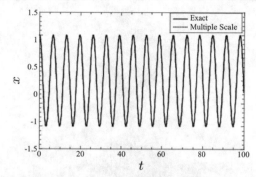

Figure 4.4: Exact solution (solid line) vs. approximate solution by multiple scale (dashed line). ($\theta_{\max} = 45°$.)

COFFEE BREAK: MULTIPLE SCALE METHOD

There is a science fiction story that I always remember whenever I think about the multiple time scale method. It was a very short science fiction (SF) story that was published in a newspaper for junior high-school students in Japan more than 50 years ago. Since it was a long time ago, I do not remember even the title of the story, but I remember that the outline of the story was as follows.

An archaeologist found the ruins of an ancient civilization in a desert. There stood a huge stone statue of guardian guarding the holy temple. The archaeologist broke a part of the toe of the statue with a hammer, and brought home a piece of stone for further examination. Unfortunately, a war broke out shortly after this event and the investigation was interrupted. When the archaeologist visited the site again several years after the war ended, he saw that the upper body of the stone statue, which used to be upright, was slightly bowed forward as if the guardian was trying to reach toward the fingertip of the foot that the archaeologist had broken with a hammer several years ago. The story ends with just this. However, the story seems to be based on the concept that various phenomena of this world are progressing in parallel at completely different time scales from each other, which is exactly the same as the basic concept of the multiple scale method, and I believe that this story is one of the masterpieces of short SF.

There is a story similar to this, not in the SF world, but a real fluid mechanical phenomenon called **pitch-drop experiment**. The pitch is a black material that remains after distillation of petroleum and other materials from crude oil. It is liquid at high temperature but it is solid at normal temperature. It suffices if you can imagine the asphalt used for road pavement. In 1927, Prof. Parnell of the University of Queensland in Australia poured pitch that had been heated to a high temperature into a funnel that was closed at the outlet, and left it at room temperature for about 3 years. Of course, at this point, the pitch was completely solidified, and if it was hit with a hammer, it would be shattered. Prof. Parnell then cut the mouth of the funnel that had been closed and set the funnel upright so that the pitch could flow out. Then eight

Figure 4.5: The pitch-drop experiment.

years later, in 1938, one drop of pitch dropped from the mouth of the funnel. This experiment is still ongoing 90 years after the start of the experiment, and 8 pitch drops have dropped so far. A mass of pitch, that appears to be completely solid and stationary on the time scale of our daily life, also flows exactly like a liquid on a scale longer than a few decades. Incidentally, this experiment is registered in the Guinness Book of Records as the longest continuous experiment in the world.

4.5   REFERENCES

[1] C. M. Bender and S. A. Orszag. *Advanced Mathematical Methods for Scientists and Engineers*. Springer, 1999. DOI: 10.1007/978-1-4757-3069-2 76

[2] A. Jeffrey and T. Kawahara. *Asymptotic Methods of Nonlinear Wave Theory*. Pitman, London, 1982. 76

[3] A. H. Nayfeh. *Introduction to Perturbation Techniques*. Wiley, 1981. 76

CHAPTER 5

KdV Equation: Effect of Dispersion

In Chapter 1, we studied that in the nonlinear wave equation $u_t + c(u)u_x = 0$, the propagation velocity $c(u)$ depends on the dependent variable u, so the waveform steepens with propagation and the solution breaks down in finite time. In the case of surface waves, this occurs when introducing nonlinear effects to waves whose wavelength is very long compared to the water depth. (See Appendices B and F for details.) On the other hand, we also studied in Chapter 3 that the surface wave is a dispersive wave, except for the long wave limit, and that the propagation velocity is different depending on the wavelength. In this chapter, we investigate the nature of the waves when both the effects of nonlinearity and dispersion are considered simultaneously. We will learn that a solitary wave that propagates extremely stably becomes possible as a result of the competition between these two effects.

5.1 KDV EQUATION AND ITS INTUITIVE DERIVATION

When a wave propagates in the positive x direction and enters into a quiescent water area with depth h which is very shallow compared to the wavelength of the wave, the surface displacement $\eta(x, t)$ is governed by the nonlinear wave equation[1]

$$\frac{\partial \eta}{\partial t} + \left(3\sqrt{g(h + \eta)} - 2\sqrt{gh}\right)\frac{\partial \eta}{\partial x} = 0. \tag{5.1}$$

As the propagation speed is an increasing function of η, the larger the η (higher water surface), the faster it travels. For this reason, as we saw in Chapter 1, the portion of $\eta_x < 0$ of the waveform leans forward more and more over time, and finally the surface displacement (hence also the flow velocity u) becomes a multi-valued function of x, which is physically unacceptable. (5.1) is an approximate equation derived under the assumption of "long wave," that is, the typical length of the spatial change of $\eta(x, t)$ is much longer than the water depth h. However, when the waveform leans forward and steepens due to nonlinear effects, the assumption of long wave is violated there. In this sense, (5.1) is an equation which involves a self-contradiction that the foundation of its own establishment is broken due to the nonlinearity that it has.

---

[1]The derivation of this equation is given in Appendix B.

On the other hand, as shown in Chapter 3, the gravity wave on water surface is a dispersive wave whose wave speed differs depending on the wavelength, and its linear dispersion relation is given by

$$\omega = \sqrt{gk \tanh kh}. \tag{5.2}$$

For "perfectly long wave" (i.e., shallow water limit $kh \to 0$) whose wavelength $\lambda = 2\pi/k$ is very long compared to the water depth h, since $\tanh kh/kh \to 1$, $\omega \to \sqrt{gh}\, k$ (wave velocity $c = \omega/k \to \sqrt{gh}$ = constant), and the dispersion disappears. Therefore, the dispersive nature of water wave is not reflected at all in (5.1) which has been derived assuming a perfectly long wave. However, with the steepening of the waveform brought about by the nonlinear effect, the state of "perfectly long wave" cannot persist forever, and a situation arises in which the dispersive character of the wave must necessarily be taken into consideration.

It is of course possible to systematically derive a new equation that takes into account the effect of dispersion starting from the system (3.27) of basic equations of water waves. However, this method is rather cumbersome to carry out, so let's turn it into Appendix F, and here we will consider incorporating the effect of dispersion in (5.1) in a more intuitive way as follows. Taylor expansion of the linear dispersion relation (5.2) of gravity wave around the long wave limit $kh = 0$ gives

$$\omega = \sqrt{gh}\, k - \frac{1}{6}\sqrt{gh}h^2 k^3 + O(kh)^5. \tag{5.3}$$

Here the part after the second term on the right side brings about the dispersion. As long as we are considering the modification of (5.1), the waves we are targeting here are long waves to some extent, and in that sense kh is a small quantity. Then, each term after the second term on the right side of (5.3) becomes smaller toward the back, so it should be possible to capture the most important part of the dispersion effect for long waves only by considering the second term. Recalling the correspondence

$$\frac{\partial}{\partial t} \longrightarrow -i\omega, \qquad \frac{\partial}{\partial x} \longrightarrow ik \tag{5.4}$$

that holds for the linear sinusoidal wave $a\, e^{i(kx-\omega t)}$ as we saw in Section 3.1, we can see that it is sufficient to add a term $(\sqrt{gh}\, h^2/6)\eta_{xxx}$ in order to incorporate the effect of the second term of the dispersion relation (5.3) into (5.1). Thus, for long waves propagating in the x direction, the equation

$$\frac{\partial \eta}{\partial t} + \left(3\sqrt{g(h+\eta)} - 2\sqrt{gh}\right)\frac{\partial \eta}{\partial x} + \frac{1}{6}\sqrt{gh}h^2 \frac{\partial^3 \eta}{\partial x^3} = 0 \tag{5.5}$$

can be obtained as an equation that incorporates the effects of nonlinearity and dispersion simultaneously.

However, there is one point to be considered here. For the derivation of (5.1), it is assumed that the wave is a perfectly long wave hence kh is very small, but it is not assumed that nonlinearity η/h is small. However, since we incorporated the effect of dispersion into (5.1) in the form of the additional term $(\sqrt{gh}h^2/6)\eta_{xxx}$ based on the linear dispersion relation, the equation has rationality only for waves close to linear. Therefore, considering that

$$3\sqrt{g(h+\eta)} - 2\sqrt{gh} = \sqrt{gh}\left(3\sqrt{1+\frac{\eta}{h}} - 2\right) = \sqrt{gh}\left\{1 + \frac{3}{2}\frac{\eta}{h} + O\left(\frac{\eta^2}{h^2}\right)\right\} \qquad (5.6)$$

under the approximation $\eta/h \ll 1$, it is reasonable to replace the nonlinear term in (5.5) with $\sqrt{gh}(1 + 3\eta/2h)$ at the same time. Thus, the equation

$$\frac{\partial\eta}{\partial t} + \sqrt{gh}\frac{\partial\eta}{\partial x} + \frac{3}{2}\sqrt{\frac{g}{h}}\,\eta\frac{\partial\eta}{\partial x} + \frac{1}{6}\sqrt{gh}h^2\frac{\partial^3\eta}{\partial x^3} = 0 \qquad (5.7)$$

is finally obtained as an equation which governs the weak nonlinear long waves propagating in the positive x direction. Equation (5.7) was first derived by Korteweg and de Vries in 1895 [12], and is called the **Korteweg-de Vries equation** or **KdV equation** for short.

The original analysis of Korteweg and de Vries also considered the effect of surface tension. In that case, if we use the expansion

$$\omega = \sqrt{\left(gk + \frac{\tau}{\rho}k^3\right)\tanh kh} = \sqrt{\left\{\frac{g}{h}(kh) + \frac{\tau}{\rho h^3}(kh)^3\right\}\left\{(kh) - \frac{1}{3}(kh)^3 + \cdots\right\}}$$

$$= \sqrt{gh}\,k - \frac{\sqrt{gh}}{2}\left(\frac{h^2}{3} - \frac{\tau}{\rho g}\right)k^3 + \cdots \qquad (5.8)$$

in kh of the linear dispersion relation of the capillary-gravity wave (3.35), the KdV equation can be derived in the same way as follows:

$$\frac{\partial\eta}{\partial t} + \sqrt{gh}\frac{\partial\eta}{\partial x} + \frac{3}{2}\sqrt{\frac{g}{h}}\,\eta\frac{\partial\eta}{\partial x} + \frac{\sqrt{gh}}{2}\left(\frac{h^2}{3} - \frac{\tau}{\rho g}\right)\frac{\partial^3\eta}{\partial x^3} = 0. \qquad (5.9)$$

As shown by (5.8), if $\frac{h^2}{3} - \frac{\tau}{\rho g} > 0$, including the case of perfect gravity waves ($\tau = 0$), the effect of dispersion (i.e., the effect that the wavelength is not infinite) works to lower the frequency (and hence the propagation velocity) and is called "negative dispersion." On the other hand, when the effect of surface tension is large so that $\frac{h^2}{3} - \frac{\tau}{\rho g} < 0$, the effect of dispersion works to increase the frequency and the propagation velocity, and is called "positive dispersion."

In the above, we used the water wave as an example and derived the KdV equation in an intuitive way. However, as its derivation process shows, any system, regardless of what physical phenomena it describes, can be approximated by the KdV equation

$$\frac{\partial u}{\partial t} + c_0\frac{\partial u}{\partial x} + \alpha u\frac{\partial u}{\partial x} + \beta\frac{\partial^3 u}{\partial x^3} = 0 \qquad (5.10)$$

in the weakly nonlinear and weak dispersion limit, as long as the system satisfies the following three requirements.

1. Dispersion disappears at the long wave limit $k \to 0$, and the wave speed of the linear sinusoidal wave approaches some constant value c_0.

2. The velocity $c(k)$ of the linear sinusoidal wave is expanded as $c = c_0 - \beta k^2 + \cdots$ in the vicinity of $k = 0$.

3. The change of wave velocity due to nonlinearity can be approximated by the first-order term αu of the disturbance $u(x, t)$.

Therefore, the KdV equation actually appears in common to describe various wave phenomena such as ion sound waves and magnetic sound waves in plasma, waves in gas-liquid multiphase fluid, and waves traveling in nonlinear lattices, etc. Thus, the KdV equation is one of the very general and important nonlinear wave equations.

In the following, we introduce a coordinate system that translates with c_0 and write the KdV equation as

$$\frac{\partial u}{\partial t} + \alpha u \frac{\partial u}{\partial x} + \beta \frac{\partial^3 u}{\partial x^3} = 0. \tag{5.11}$$

5.2   SOLITARY WAVE SOLUTION: BALANCE BETWEEN NONLINEARITY AND DISPERSION

Equation (5.11) has the following steady traveling wave solution that propagates at a constant speed without changing its form:

$$u(x, t) = a \operatorname{sech}^2 \left[\sqrt{\frac{\alpha a}{12 \beta}} (x - ct - x_0) \right], \qquad c = \frac{\alpha}{3} a. \tag{5.12}$$

This is called the **solitary wave solution** of the KdV equation and has the following characteristics.

- It has a pulse shape that is symmetrical about the center of the wave and asymptotically approaches $u = 0$ as $x \to \pm\infty$. (This is the reason why it is called solitary wave.) [2]

- The propagation velocity of the solitary wave (more precisely, the deviation from the linear long wave velocity c_0) is proportional to the amplitude a. In the case of $\alpha > 0$ ($\alpha < 0$), larger solitary waves travel faster (slower).

- The width of the wave is inversely proportional to $\sqrt{|a|}$, and the larger the solitary wave, the thinner it is.

---

[2]The transformation $u \longrightarrow u + u_\infty$, $\quad c \longrightarrow c + \alpha u_\infty$ also gives solutions that asymptote to any non-zero constant u_∞.

If there is no dispersion term u_{xxx} in (5.11), any waveform that is input initially steepens forward (when $\alpha > 0$) or backward ($\alpha < 0$) as time, and cannot propagate with maintaining a constant waveform. Also, if there is no nonlinear term in (5.11), the various wavelength components that make up the initial waveform propagate at different speeds due to the dispersion term, and it is not possible to propagate with a constant waveform as well. Therefore, the solitary wave solution (5.12) is realized only by the coexistence of the nonlinear effect and the dispersion effect.

In order for a solitary wave solution (5.12) to exist, $\alpha a/\beta$ in the square root must be positive. This is a good indication that the solitary wave solution is indeed built on a balance between nonlinear and dispersive effects. For example, in the case of gravity waves (5.7), $\alpha > 0$, $\beta > 0$, and this condition requires that $a > 0$, that is, the solitary wave is always a swell of the water surface as shown in Fig. 5.1a. When $\alpha > 0$, the nonlinear effect makes the higher part of the wave to propagate faster, so the waveform tends to lean forward, resulting in generation of many short wavelength components. However, if $\beta > 0$, i.e., there is negative dispersion, these short wavelength components can propagate only slower than long waves, and there is an effect of pulling back the forward-leaning waveform due to nonlinearity. In the situation where $\alpha > 0$, $\beta > 0$, if $a < 0$, that is, if there is a dip instead of a swell on the water surface, both the nonlinearity and the dispersion become the effect of pulling back the waveform, and the wave cannot travel while maintaining a constant waveform. Therefore, physical consideration also tells that only $a > 0$, that is, upward convex solitary waves should exist in the case of gravity waves ($\alpha > 0$, $\beta > 0$). On the other hand, when the surface tension is superior to gravity, $\alpha > 0$, $\beta < 0$, the competition and balance between the nonlinear effect and the dispersion effect becomes possible only in the case of $a < 0$, and as a result, only solitary waves with depressed water surface exist.

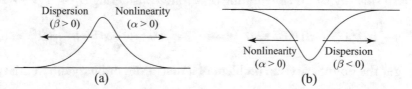

Figure 5.1: Competition of nonlinearity and dispersion in solitary wave: (a) gravity wave type and (b) capillary wave type.

The width d of the solitary wave solution (5.12) can be estimated by the reciprocal of the coefficient of x, and it can be seen that d is related to the coefficients α, β of the KdV equation and the height a of the solitary wave like $d = \sqrt{\beta/\alpha a}$. This also indicates that the solitary wave solution is realized on the balance between nonlinearity and dispersion. When the representative length of the waveform change is d, the magnitudes of the nonlinear term and the dispersion term of (5.11) can be estimated by $O(\alpha a^2/d)$ and $O(\beta a/d^3)$, respectively. In order to balance the nonlinear effect and the dispersive effect, these two terms need to be of the same size. This requires that

$$\frac{\alpha a^2}{d} \sim \frac{\beta a}{d^3} \quad \longrightarrow \quad d \sim \sqrt{\frac{\beta}{\alpha a}}, \tag{5.13}$$

which is exactly the width of the solitary wave solution (5.12). That is, the width of the solitary wave is adjusted so that the magnitudes of the nonlinear and the dispersive effects become comparable.

EXAMPLE 1: DERIVATION OF SOLITARY WAVE SOLUTION

Find the solitary wave solution (5.12) as the steady traveling wave solution of the KdV equation (5.11).

[Answer]

The problem is reduced to the boundary value problem of an ordinary differential equation in the same way as we obtained the shock wave solution of the Burgers equation as the steady traveling wave solution in Example 2 of Chapter 2. First, assuming that $u(x, t) = U(\xi)$, $\xi = x - ct$, and substituting into the KdV equation (5.11), the following ordinary differential equation for $U(\xi)$ is obtained:

$$-cU' + \alpha U U' + \beta U''' = 0, \tag{5.14}$$

where $'$ represents the derivative with respect to ξ. Integrating this once with taking into consideration the boundary condition $U \to 0$ ($x \to \pm\infty$) gives

$$-cU + \frac{1}{2}\alpha U^2 + \beta U'' = 0. \tag{5.15}$$

Multiplying both sides by U' and integrating once, and considering $U \to 0$ ($x \to \pm\infty$), we get

$$-cUU' + \frac{1}{2}\alpha U^2 U' + \beta U'' U' = 0 \quad \longrightarrow \quad -\frac{1}{2}cU^2 + \frac{1}{6}\alpha U^3 + \frac{1}{2}\beta U'^2 = 0, \tag{5.16}$$

and finally we get the boundary value problem of a first-order ordinary differential equation

$$\left(\frac{dU}{d\xi}\right)^2 = \frac{\alpha}{3\beta}U^2\left(\frac{3c}{\alpha} - U\right), \quad U \to 0 \ (\xi \to \pm\infty). \tag{5.17}$$

It is left for the reader to solve this, and here we only confirm that the solitary wave solution (5.12) satisfies this equation. First of all, from

$$\operatorname{sech} x = (\cosh x)^{-1} \quad \longrightarrow \quad (\operatorname{sech} x)' = -\operatorname{sech} x \tanh x, \tag{5.18}$$

$y(x) = \operatorname{sech}^2 x$ satisfies the equation

$$\left(\frac{dy}{dx}\right)^2 = 4y^2(1 - y), \quad y \to 0 \ (x \to \pm\infty), \tag{5.19}$$

and therefore $y(x) = a \operatorname{sech}^2(bx)$ satisfies

$$\left(\frac{dy}{dx}\right)^2 = \frac{4b^2}{a} y^2(a - y), \quad y \to 0 \ (x \to \pm\infty). \tag{5.20}$$

Comparing the coefficients of this equation and (5.17), it can be seen that if

$$a = \frac{3c}{\alpha}, \text{ i.e., } c = \frac{\alpha}{3}a, \quad \text{and} \quad \frac{4b^2}{a} = \frac{\alpha}{3\beta}, \text{ i.e., } b = \sqrt{\frac{\alpha a}{12\beta}}, \tag{5.21}$$

then $U(\xi) = a \operatorname{sech}^2(b\xi)$ satisfies (5.17), and gives the solitary wave solution of the KdV equation (5.11). ♣

. . . .

Although the concrete expression is not described here, Korteweg and de Vries (1895) also derived a periodic steady traveling wave solution with a fixed wavelength in addition to the solitary wave solution. The solution is expressed using Jacobi's elliptic function cn, and is called the **cnoidal wave**. The cnoidal wave agrees with the solitary wave solution at the limit of wavelength $\to \infty$ and the sinusoidal wave of the linearized KdV equation at the limit of amplitude $\to 0$. In this sense, the cnoidal wave plays a role of continuously connecting the solitary wave solution, which is a typical nonlinear wave, and the linear sinusoidal wave solution.

5.3   SOLITON: SOLITARY WAVE WITH PARTICLE NATURE

5.3.1   DISCOVERY OF SOLITON

It was the work of Zabusky and Kruskal (1965) [21] that made the KdV equation, which had already been derived in 1895, famous again. They performed numerical simulations to solve the initial and boundary value problem of the KdV equation as follows[3]:

$$\frac{\partial u}{\partial t} + u\frac{\partial u}{\partial x} + \delta^2 \frac{\partial^3 u}{\partial x^3} = 0, \quad \delta = 0.022, \quad u(x,0) = \cos \pi x, \quad u(0,t) = u(2,t). \tag{5.22}$$

As a result, they found following properties.

1. The initial waveform splits into multiple pulses under the influence of the dispersion term after steepening due to the nonlinear term.

2. These pulses can be considered as solitary wave solutions of the KdV equation from their waveforms and from the relationship between wave height, width, and propagation velocity.

[3]In the background of their numerical study of KdV equation, there was a wider awareness on nonlinear physical problem, which is related to the basis of statistical mechanics. See Appendix G for this point.

3. Since these solitary waves have different propagation speeds, they collide with each other under the periodic boundary condition. However, in each case, it reappears after a collision with almost no change in size or shape.

Figure 5.2 shows the result of the numerical simulation almost the same as theirs. In the right figure, the value of u is expressed in grayscale, and the whitish lines represent the movement of the peak positions of the solitary waves formed by the split. It can be seen that the solitary waves move with holding their identity while colliding. Zabusky and Kruskal named this solitary wave **soliton** because the solitary wave of the KdV equation is thus extremely stable to the interaction, as if it behaved like a particle.[4]

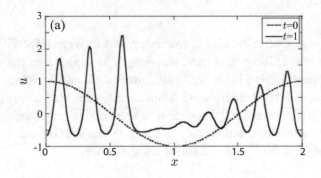
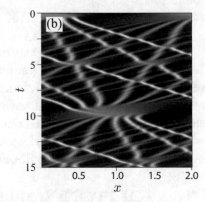

Figure 5.2: Numerical simulation of Zabusky–Kruskal (1965) [21]: (a) the profile of u at $t = 0$ and $t = 1$ and (b) trajectories of the solitary waves. The white part has a larger value of u.

In a linear system, such particle behavior is not surprising at all. In linear systems, superposition of solutions is allowed. Therefore, if there were two solitary wave solutions $u_1(x,t)$, $u_2(x,t)$ with different propagation speeds, and if we employed $u(x,0) = u_1(x,0) + u_2(x,0)$ as the initial condition with the initial positions being adjusted so that the faster of the two comes behind, then it is obvious that the faster solitary wave behind would overtake the slower solitary wave ahead with time without loosing its identity.

The reason that the numerical result of Zabusky and Kruskal (1965) was so surprising is that the KdV equation is a nonlinear equation for which the superposition is not allowed. In the KdV equation, $u_1(x,t) + u_2(x,t)$ cannot be a solution even if $u_1(x,t)$ and $u_2(x,t)$ are solutions. As we saw in the previous section, the solitary wave solution of the KdV equation is an essentially nonlinear solution which is realized on the balance between the nonlinear effect and the dispersive effect. It is very strange that such nonlinear solutions are (looks like) passing with each other without any interaction. However, as seen coming up, the solitary waves are not

---

[4]Like protons, electrons, etc., the suffix "–on" gives the impression of "particles."

actually just passing through each other, but exert a strong and complicated nonlinear interaction with each other at the time when they are close.

In Section 1.2 of Chapter 1, we studied that when the functions T, X of the dependent variable u satisfy

$$\frac{\partial T}{\partial t} + \frac{\partial X}{\partial x} = 0, \tag{5.23}$$

this expresses a conservation law of certain physical quantity such that T and X are its density and flux, respectively. If X becomes 0 as $|x| \to \infty$, (5.23) means that $\frac{d}{dt} \int_{-\infty}^{\infty} T \, dx = 0$, that is, $\int_{-\infty}^{\infty} T \, dx$ is a conserved quantity. In fact, it is known that the KdV equation has an infinite number of conservation laws like (5.23) [15]. Specifically, for the KdV equation of the form (5.11), the first three are given as follows:

$$\frac{\partial u}{\partial t} + \frac{\partial}{\partial x} \left(\frac{1}{2} \alpha u^2 + \beta u_{xx} \right) = 0, \tag{5.24a}$$

$$\frac{\partial u^2}{\partial t} + \frac{\partial}{\partial x} \left(\frac{2}{3} \alpha u^3 + 2\beta u u_{xx} - \beta u_x^2 \right) = 0, \tag{5.24b}$$

$$\frac{\partial}{\partial t} \left(\alpha u^3 - 3\beta u_x^2 \right)$$
$$+ \frac{\partial}{\partial x} \left(\frac{3}{4} \alpha^2 u^4 + 3\alpha\beta u^2 u_{xx} - 6\alpha\beta u u_x^2 - 6\beta^2 u_x u_{xxx} + 3\beta^2 u_{xx}^2 \right) = 0. \tag{5.24c}$$

It is known that there is a close relationship between the particle-like stability and self-holding ability of the solitons and the fact that the KdV equation has an infinite number of conservation laws. Also, apart from the types of conservation laws for which the density is given by polynomials of u and its x derivatives as above, the following is known to hold as a conservation law that depends explicitly on x:

$$\frac{d}{dt} \int_{-\infty}^{\infty} xu \, dx = \text{constant}, \tag{5.25}$$

which means that the propagation velocity of the "center of mass" of u is constant.

5.3.2   INVERSE SCATTERING METHOD: EXACT SOLUTION OF KDV EQUATION

Immediately after the impact of soliton discovery by Zabusky and Kruskal [21], Gardner et al. [6, 7] developed an innovative method to analytically solve the initial value problem of the KdV equation for any initial conditions that decays fast enough in $|x| \to \infty$. In this method, the solution $u(x, t)$ of the KdV equation

$$\frac{\partial u}{\partial t} - 6u \frac{\partial u}{\partial x} + \frac{\partial^3 u}{\partial x^3} = 0 \tag{5.26}$$

is identified with the potential $u(x)$ of the time-independent Schrödinger equation

$$-\frac{d^2\phi}{dx^2} + u\phi = \lambda\phi \qquad (5.27)$$

in quantum mechanics.[5]

If the potential $u(x)$ decays at $|x| \to \infty$, (5.27) becomes

$$-\frac{d^2\phi}{dx^2} = \lambda\phi \qquad (5.28)$$

at $|x| \to \infty$, so for states with energy level $\lambda > 0$ and can be written as $\lambda = k^2$ ($k > 0$), the behavior of $\phi(x)$ for $|x| \to \infty$ is a linear combination of $e^{\pm ikx}$. By properly normalizing, we can construct a solution that satisfies the boundary condition

$$\phi(x) \sim \begin{cases} e^{-ikx} + R(k)e^{ikx} & (x \to +\infty) \\ T(k)e^{-ikx} & (x \to -\infty). \end{cases} \qquad (5.29)$$

This is called the wave function of scattering state. This solution corresponds to the situation that a material wave of amplitude 1 is incident from $x \to +\infty$, of which $R(k)e^{ikx}$ is reflected by the potential $u(x)$ and $T(k)e^{-ikx}$ is transmitted. Therefore, $R(k)$ and $T(k)$ are called reflection coefficient and transmission coefficient, respectively, and $|R(k)|^2 + |T(k)|^2 = 1$ holds from the conservation of probability.

On the other hand, for the state with $\lambda < 0$ and can be written as $\lambda = -\kappa^2$ ($\kappa > 0$), the behavior of $\phi(x)$ at $|x| \to \infty$ is a linear combination of $e^{\pm \kappa x}$, and it must be like

$$\phi(x) \sim \begin{cases} c(\kappa)e^{-\kappa x} & (x \to +\infty) \\ d(\kappa)e^{\kappa x} & (x \to -\infty) \end{cases} \qquad (5.30)$$

in order for $\phi(x)$ not to diverge at $|x| \to \infty$. The existence of such exponentially decaying wave functions in both $x \to \pm\infty$ directions is possible only for finite number of specific energy levels $\lambda_n = -\kappa_n^2$ ($n = 1, 2, \ldots, N$) determined from the potential $u(x)$. In addition, when the normalization condition $\int_{-\infty}^{\infty} |\phi(x)|^2 \, dx = 1$ is requested, the coefficients c_n ($= c(\kappa_n)$) are determined, and corresponding d_n ($= d(\kappa_n)$) are also determined automatically. The $\phi(x, t)$ thus normalized is called the wave function of bound state (Fig. 5.3).

The combination of the reflection coefficient $R(k)$ of the scattering state and κ_n ($n = 1, 2, \ldots, N$) and the normalization coefficient c_n ($n = 1, 2, \ldots, N$) of the bound state is called the scattering data The problem of finding these scattering data for a given potential $u(x)$ is called

---

[5]Quantum mechanics is the dynamics that governs physical phenomena in the micro world such as molecules and atoms, and protons, neutrons and electrons that constitute them. And the Schrödinger equation (5.27) is an equation that the steady-state wave function $\phi(x)$ of energy level λ should satisfy, and the existence probability of the micro particle is proportional to $|\phi(x)|^2$.

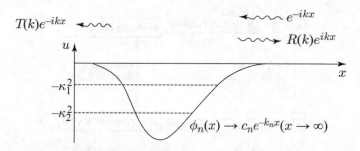

Figure 5.3: Scattering state and bound state of a potential $u(x)$.

the scattering problem. Conversely, it is also known that the potential $u(x)$ can be determined from the scattering data alone, and a specific method for obtaining $u(x)$ from the scattering data has also been developed.[6] The problem of determining the potential from scattering data in this way is called the inverse scattering problem.

Gardner et al. investigated the time evolution of each of the scattering data when the potential $u(x)$ of the Schrödinger equation (5.27) changes with time according to the KdV equation (5.26). As a result, they found that they can be expressed in very simple forms that do not include $u(x, t)$, as follows:

$$\frac{d\kappa_n}{dt} = 0 \quad \longrightarrow \quad \kappa_n(t) = \kappa_n(0), \tag{5.31a}$$

$$\frac{dc_n}{dt} = 4\kappa_n^3 c_n \quad \longrightarrow \quad c_n(t) = c_n(0)\, e^{4\kappa_n^3 t}, \tag{5.31b}$$

$$\frac{dR(k)}{dt} = 8ik^3 R(k) \quad \longrightarrow \quad R(k, t) = R(k, 0)\, e^{8ik^3 t}. \tag{5.31c}$$

Based on the above results, Gardner et al. proposed the following method to analytically solve the initial value problem of the KdV equation.

1. First, solve the scattering problem of the Schrödinger equation (5.27) with the initial waveform $u(x, 0)$ of the KdV equation as the potential, and find the scattering data $\{R(k, 0), \kappa_n(0), c_n(0)\}$ for $u(x, 0)$.

2. Find the scattering data $\{R(k, t), \kappa_n(t), c_n(t)\}$ at time t when the potential $u(x)$ changes with time according to the KdV equation from (5.31).

3. Solve the inverse scattering problem for the scattering data at time t and find the corresponding potential $u(x, t)$, i.e., the solution of the KdV equation at time t.

This method of solution of the KdV equation is called the **inverse scattering method**. The procedure of the inverse scattering method is shown graphically in Fig. 5.4.

[6]Although not specifically described here, it results in solving a linear integral equation called the Gel'fand–Levitan–Marchenko equation.

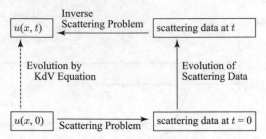

Figure 5.4: Solution procedure of the inverse scattering method.

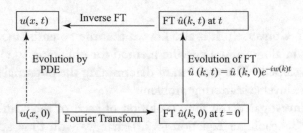

Figure 5.5: Solution procedure of linear problem.

This method is similar to the method of solving an initial value problems of a linear partial differential equation with constant coefficients using Fourier transform (See Fig. 5.5). Let

$$\frac{\partial u}{\partial t} + \mathcal{L}(u) = 0 \tag{5.32}$$

be a constant coefficient linear PDE to be solved. Here, $\mathcal{L}(u)$ is linear in $u(x,t)$ and includes u and its partial derivatives with respect to x. Let $\omega = \omega(k)$ be the linear dispersion relation obtained by substituting $a\, e^{i(kx-\omega t)}$ into (5.32). When the initial waveform $u(x,0)$ is represented by Fourier transform as

$$u(x,0) = \int_{-\infty}^{\infty} \hat{u}(k,0)\, e^{ikx}\, dk, \tag{5.33}$$

the waveform $u(x,t)$ at an arbitrary time t is given by

$$u(x,t) = \int_{-\infty}^{\infty} \hat{u}(k,0)\, e^{i[kx-\omega(k)t]}\, dk = \int_{-\infty}^{\infty} \left[\hat{u}(k,0)e^{-i\omega(k)t} \right] e^{ikx}\, dk, \tag{5.34}$$

That is, the initial value problem of (5.32) can be solved by the following procedure.

1. Find the Fourier transform $\hat{u}(k,0)$ of the initial waveform $u(x,0)$.

2. Find the Fourier transform $\hat{u}(k,t)$ at t by $\hat{u}(k,0)\exp[-i\omega(k)t]$.

3. Find $u(x,t)$ by inverse Fourier transform of $\hat{u}(k,t)$.

Looking at the relationship with the inverse scattering method, the Fourier transform and the inverse Fourier transform correspond to the process of solving the scattering problem and the inverse scattering problem, respectively, the Fourier transform $\hat{u}(k, 0)$ corresponds to the scattering data, and the process of updating $\hat{u}(k, t)$ using the dispersion relation as $\hat{u}(k, 0) \exp[-i\omega(k)t]$ corresponds to the process of updating the scattering data using (5.31).

A particularly important point in the connection between the KdV equation (5.26) and the Schrödinger equation (5.27) is that as long as $u(x, t)$ changes according to the KdV equation, no matter how much the waveform changes, the energy level $-\kappa_n^2$ of the bound state of the Schrödinger equation does not change as shown in (5.31a). According to the inverse scattering method, each of the N bound states of the Schrödinger equation (5.27) corresponds to one solitary wave

$$u_i(x, t) = -2\kappa_i^2 \operatorname{sech}^2 \left\{ \kappa_i(x - 4\kappa_i^2 t - x_i) \right\} \quad (i = 1, \ldots, N) \tag{5.35}$$

of the KdV equation, and it can be said that the invariance of this energy level supports the remarkable stability of KdV solitons.

The KdV equation is an equation originally derived as an equation describing water waves with long wavelength. On the other hand, the Schrödinger equation is a governing equation in quantum mechanics targeting atomic and elementary particle level motion. The keen insight (inspiration) of Gardner et al. that makes it possible to develop an innovative solution method for the initial value problem of nonlinear PDE by connecting two equations that have been known in completely different contexts may be a good example of "serendipity."

For more details of the inverse scattering method, see [1] and [19], for example.

5.3.3   SOLITON INTERACTION

For the KdV equation, an analytical solution called "N-soliton solution" that describes the interaction of N solitons is known. Here, as a typical example of soliton interaction, we will show only the 2-soliton solution expressing overtaking of two solitons. The two soliton solution $u(x, t)$ of the KdV equation (5.26) is represented as

$$u = -2(\ln f)_{xx}, \qquad f = 1 + E_1 + E_2 + \left(\frac{\kappa_1 - \kappa_2}{\kappa_1 + \kappa_2}\right)^2 E_1 E_2,$$

$$E_i = \exp[2\kappa_i(x - 4\kappa_i^2 t - x_i)] \quad (i = 1, 2). \tag{5.36}$$

Here, if $\kappa_2 = 0$, it results in the 1-soliton solution (5.35) corresponding to κ_1. As an example, the case of $\kappa_1 = 1$, $\kappa_2 = 2$ is shown in Fig. 5.6. From (5.35), this solution shows that the soliton with height 8 and velocity 16 overtakes the soliton with height 2 and velocity 4.

At first glance, the high and fast soliton seems to overtake the low and slow soliton as if they were linear waves and without any interaction. However, looking closely, we can find evidences that there is significant nonlinear interactions between them. One evidence is in the waveform when two solitons overlap. If it is a simple linear wave interaction, the height of the

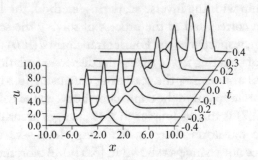

Figure 5.6: 2-soliton solution at various t. ($\kappa_1 = 1$, $\kappa_2 = 2$.)

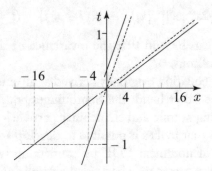

Figure 5.7: Phase shift in soliton interaction.

waveform that is generated when the two overlap should be 10 ($= 8 + 2$), but the height of the waveform at that time is only 6 as shown in Fig. 5.6. This alone shows that the two solitary waves do not simply overlap.

Another evidence is in the phase shift. Figure 5.7 shows the trajectories of the two solitons in the xt-plane. As you can see from the figure, the position of the two solitons are different from that when there is no interaction between them (dotted line). It can be seen that the higher soliton is pushed forward, while the lower soliton is pulled backward.

Although in the case treated here the waveform has only one peak when the two solitons overlap, it is known that the behavior of the waveform when two solitons approach is classified into the following two types depending on the amplitude ratio. In the case of $\kappa_2^2/\kappa_1^2 > (3 + \sqrt{5})/2 \approx 2.62$, the smaller soliton is swallowed by a larger soliton, and the waveform temporarily becomes one peak as in the case shown in Fig. 5.6. On the other hand, in the case of $1 < \kappa_2^2/\kappa_1^2 < (3 + \sqrt{5})/2$, an "exchange of roles" takes place between the solitons and the waveform always keeps to have two peaks. Interestingly enough, this result was derived using only the first three conservation laws (5.24) of the KdV equation, without knowing the exact 2-soliton solution [13].

As a method of finding soliton solutions, **Hirota's direct method** is also well known [9]. This method enables us to obtain the N-soliton solution without using the inverse scattering method.

EXAMPLE 3: NUMERICAL SIMULATION OF KDV EQUATION

Reproduce the overtaking of two solitons shown in Fig. 5.6 by direct numerical simulation of the KdV equation (5.26).

[Answer]

Zabusky–Kruskal (1965) used the following finite-difference scheme:

$$\frac{\partial u}{\partial t} - 6u\frac{\partial u}{\partial x} + \frac{\partial^3 u}{\partial x^3} = 0,$$

$$\longrightarrow \quad \frac{u_i^{j+1} - u_i^{j-1}}{2\Delta t} - 6\left(\frac{u_{i+1}^j + u_i^j + u_{i-1}^j}{3}\right)\left(\frac{u_{i+1}^j - u_{i-1}^j}{2\Delta x}\right)$$

$$+ \frac{u_{i+2}^j - 2u_{i+1}^j + 2u_{i-1}^j + u_{i-2}^j}{2(\Delta x)^3} = 0,$$

$$\longrightarrow \quad u_i^{j+1} = u_i^{j-1} + \frac{2\Delta t}{\Delta x}\left(u_{i+1}^j + u_i^j + u_{i-1}^j\right)\left(u_{i+1}^j - u_{i-1}^j\right)$$

$$- \frac{\Delta t}{(\Delta x)^3}\left(u_{i+2}^j - 2u_{i+1}^j + 2u_{i-1}^j - u_{i-2}^j\right), \qquad (5.37)$$

where $u_i^j = u(i\Delta x, j\Delta t)$. Since this scheme approximates the time derivative with a central difference, not only the values in jth steps and but also $(j-1)$th steps are required to calculate the value of $(j+1)$th step.[7] Therefore, even when the initial condition is given, it cannot start by itself. However, for example, the forward difference (Euler's method) may be used to calculate the first step, and then (5.37) may be used after the second step. If it is bothersome, there would be no particular problem even if you use Euler's method not only to obtain the second step but to also calculate all the steps after the second step.[8] For explicit difference schemes for the KdV equation, including the Zabusky–Kruskal scheme, see, for example, [18].

5.3.4   APPLICATION OF SOLITON THEORY TO WATER WAVES

The KdV equation (5.7) for the gravity waves on water surface can be converted to the standard form (5.26) by the variable transformation

$$\tau = \frac{1}{6}\sqrt{\frac{g}{h}}\,t, \quad \xi = \frac{(x - \sqrt{gh}\,t)}{h}, \quad u = -\frac{3\eta}{2h}. \qquad (5.38)$$

---

[7]Such a scheme is often called the "leap-frog method."
[8]Actually, in the numerical simulation shown in Fig. 5.2, all time evolution is performed by this method.

Using this correspondence, the main properties of the solution of the KdV equation that can be obtained from the theory of the inverse scattering method can be expressed in terms of water waves as follows.

1. Any initial waveform that decays fast enough as $|x| \to \infty$ splits into stationary solitary wave solutions (solitons) and dispersive wave trains (tails).

2. The amplitude of the tail decays with time like $t^{-1/3}$.

3. If the area of the initial waveform $\int_{-\infty}^{\infty} \eta(x, 0)\, dx$ is positive, at least one soliton appears.

4. No soliton appears if $\eta(x, 0) < 0$ for all x.

5. The number N of solitons emerging from the initial waveform $\eta(x, 0)$ is given by the number of bound states of the Schrödinger equation (5.27) with $u(x) = -3\eta(x, 0)/2h$ as the potential, and the wave height a_n ($n = 1, \ldots, N$) of each solitary wave is given by $a_n = 4\kappa_n^2 h/3$ by the energy level $-\kappa_n^2$ of the corresponding bound state.

The time required for the initial waveform to split into solitons can be roughly estimated as follows. Let the maximum value of the initial waveform $\eta(x, 0)$ and the typical length (width) be η_0 and l, respectively. In this case, considering (5.38), the maximum depth of the potential of the corresponding Schrödinger equation is $-3\eta_0/2h$, and the energy level of the bound state never falls below this value. This means that the maximum wave height of the soliton that can appear from this initial waveform does not exceed $2\eta_0$, so that its speed does not exceed $\sqrt{gh}(1 + \eta_0/h)$. On the other hand, since the lower limit of possible soliton speeds is $\sqrt{gh}$, a measure of the speed difference Δc between the fastest and the slowest solitons is given by $\sqrt{gh}\,(\eta_0/h)$. The measure of the time t_s required for soliton splitting can be estimated by the time when the distance between the fastest soliton and the slowest soliton becomes about the width l of the initial waveform. So, for t_s and the propagation distance d_s required for soliton splitting, we obtain the following estimate:

$$t_s = l/\Delta c = \sqrt{h/g}\,(l/\eta_0), \qquad d_s = \sqrt{gh}\,t_s = lh/\eta_0. \tag{5.39}$$

d_s gives a measure of the required length of the water tank when reproducing soliton splitting in a wave tank experiment.

The behavior of the solution of the KdV equation predicted by the inverse scattering method is compared with the results of the wave tank experiment, and it is reported that the agreement between the two regarding the number of appearing solitons and their wave heights etc. is quite good, if the effect of viscous damping at the tank wall is properly taken into consideration [8]. Also, the split into solitons has actually been reported in several tsunamis, including the giant tsunami that struck the Pacific coast of Tohoku region during the 2011 Great East Japan Earthquake.

The fact that a very stable pulse structure can exist due to nonlinear effects, like KdV solitons, has not only affected water wave researches, but has had a great impact on a wide range of natural sciences. For example, there is even a somewhat intriguing attempt to explain the sustaining mechanism of Jupiter's Great Red Spot (Fig. 5.8), that has been present for many years, using the exceptional stability of the KdV soliton [14].

Figure 5.8: The KdV soliton may have something to do with Jupiter's Great Red Spot.

5.4   RELATIVES OF KDV EQUATION

As can be seen from the argument made at the beginning of this chapter when the KdV equation (5.7) is derived in the context of water waves, at the starting point of the KdV equation there is the situation of "linear long wave" which considers neither nonlinearity nor dispersion, that is, the situation where the nonlinear effects can be neglected because the amplitude is small enough, and at the same time, the dispersion effect can also be ignored because the wavelength is long enough compared to the water depth. According to the KdV equation of the form (5.10), this situation is represented by $u_t + c_0 u_x = 0$ consisting of only the first two terms. The KdV equation (5.10) is then obtained as a result of adding to this the third term to capture the weak nonlinear effect, and adding the fourth term to capture weak dispersive effect at the same time. According to the concept of this derivation, the first and the second terms of the KdV equation (5.10) are much larger than the latter two terms.

Let us consider this a little more quantitatively, in line with the KdV equation (5.7) for water waves. Let a be the representative values of $\eta(x,t)$, l be the representative length of the spatial variation of η(i.e., x-dependence), and l/c_0 ($c_0 = \sqrt{gh}$) be the representative time of temporal variation of η. If we introduce the dimensionless parameters ϵ and μ by $\epsilon = a/h$, $\mu = (h/l)^2$, the assumption of weak nonlinearity means $\epsilon \ll 1$, and the assumption of long wave corresponds to $\mu \ll 1$. When the magnitude of each term of (5.7) is estimated using these, the first and second terms are $O(c_0 a/l)$, the third term is $O(c_0 \epsilon a/l)$, and the fourth term is $O(c_0 \mu a/l)$, so the relative magnitude to the first term is $1 : 1 : \epsilon : \mu$.

It should be remembered that, in the process of deriving the KdV equation, higher-order terms in kh and hence in μ are discarded when expanding and truncating the dispersion relation around the long wavelength limit. Also, when rewriting the nonlinear terms, higher-order terms in a/h and hence those in ϵ are discarded. Therefore, the KdV equation is an approximate equation containing errors. The right side is not really zero. There should be terms of $O(\epsilon^2, \mu^2)$ or higher-order relative to the first term, but they are just not written. If we think this way, there is no problem in rewriting, for example, the dispersion term $+u_{xxx}$ of the KdV equation (5.10) to $-u_{xxt}/c_0$. From (5.10), $u_x = -u_t/c_0 + O(\epsilon, \mu)$, thus even if we replace $u_x \rightarrow -u_t/c_0$ in the dispersion term of $O(\mu)$, the newly generated error is relatively only $O[\mu \times (\epsilon, \mu)]$, and this degree of error is ignored from the beginning. Thus, in the higher-order terms (the nonlinear term and the dispersion term) of the KdV equation (5.11), some of the x-derivatives can be replaced by t-derivatives by using the lowest order relation $\partial/\partial t \sim -c_0 \partial/\partial x$ without affecting the quality of the approximation. Therefore, we can obtain the following equations having the same validity as the KdV equation:

$$u_t + c_0 u_x + \left\{ \begin{array}{c} \alpha u u_x \\ -(\alpha/c_0)u u_t \end{array} \right\} + \left\{ \begin{array}{c} \beta u_{xxx} \\ -(\beta/c_0)u_{xxt} \\ +(\beta/c_0^2)u_{xtt} \\ -(\beta/c_0^3)u_{ttt} \end{array} \right\} = 0. \tag{5.40}$$

Among these,

$$u_t + c_0 u_x + \alpha u u_x - (\beta/c_0)u_{xxt} = 0, \tag{5.41}$$

in which one of the three x derivatives in the dispersion term is replaced by a t derivative, is studied in detail by Benjamin, Bona and Mahony (1972) [3], and, by taking the first letter of the three, is called **BBM equation**, or the "Regularized Long Wave equation," or the "RLW equation" for short. The linear dispersion relation of KdV equation (5.11) is given by

$$\omega(k) = c_0 k - \beta k^3. \tag{5.42}$$

According to this, the gravity wave with k such that $k > \sqrt{c_0/\beta}$ will propagate in the negative x direction, contrary to the assumption of the equation's derivation. Moreover, there is no lower limit to the phase velocity and group velocity, and the component with a large wavenumber is transmitted at a very high speed in the $-x$ direction. On the other hand, the linear dispersion relation of the BBM equation (5.41) is given by

$$\omega(k) = \frac{c_0 k}{1 + \beta k^2/c_0}, \tag{5.43}$$

and for small k, it naturally is very close to the dispersion relation of KdV (5.42), but unlike (5.42), gives finite positive phase velocity and group velocity for all k. BBM equations also have solitary wave solution as a steady traveling wave solution. However, when the overtaking of two

BBM solitary waves is examined by numerical simulation, a small tail part is generated after overtaking, and therefore it does not seem to be a soliton in a strict sense. As the KdV equation gets in the limelight by the discovery of solitons and the success of the inverse scattering method, the BBM equation seems to have been losing its popularity. However, as discussed above, both the KdV and the BBM equations have the same value as approximations to the original physical system. If we respect the physical meaning of these equations, it is not reasonable to argue the superiority or inferiority of these equations depending on whether their solitary wave solutions exhibit soliton-like behavior. Rather, restricting the application of these equations only to those situations in which the behaviors of their solutions do not differ appreciably would be the proper way of use as approximate equations.

In addition to the equations in (5.40), many "relatives" are known for the KdV equation. Here are some typical ones.

- **2D KdV equation**

$$(u_t + c_0 u_x + \alpha u u_x + \beta u_{xxx})_x = \gamma u_{yy}. \tag{5.44}$$

An extended version of the KdV equation that also takes into account slow changes in the y direction perpendicular to the wave propagation direction x. Since it was first investigated by Kadomtsev-Petviashvili (1970) [11], it is also called the **Kadomtsev–Petviashvili equation**, or **KP equation** for short.

- **KdV Burgers equation**

$$u_t + c_0 u_x + \alpha u u_x + \beta u_{xxx} = \nu u_{xx}. \tag{5.45}$$

In addition to nonlinearity and dispersion, this equation also considers the diffusion effect.

- **Modified KdV equation**

$$u_t + c_0 u_x + \alpha u^2 u_x + \beta u_{xxx} = 0. \tag{5.46}$$

Approximate equation for wave called "Alfven wave" in plasma, for example. Unlike the KdV equation, it has a third-order nonlinear term. This equation is also known to have played an important role in the process of Gardner et al. developing the inverse scattering method for the KdV equation.

- **Benjamin–Ono equation** [2] and [16]

$$u_t + c_0 u_x + \alpha u u_x + \gamma \mathcal{H}[u_{xx}] = 0, \tag{5.47}$$

where $\mathcal{H}$ denotes the Hilbert transform defined by

$$\mathcal{H}[f(x)] = \frac{1}{\pi} \mathcal{P} \int_{-\infty}^{\infty} \frac{f(x')}{x' - x} \, dx', \tag{5.48}$$

where $\mathcal{P}$ stands for the principal value of the singular integral. Since the Hilbert transform of e^{ikx} is given by

$$\mathcal{H}\left[e^{ikx}\right] = i\, \text{sgn}(k)\, e^{ikx}, \quad (\text{sgn is the sign function taking } +1 \text{ or } -1), \tag{5.49}$$

the linear dispersion relation of the Benjamin–Ono equation (5.47) is given by

$$\omega = c_0 k - \gamma |k| k. \tag{5.50}$$

In the case where ω is approximated in the vicinity of the long wave limit $k \to 0$ by $\omega = c_0 k - \gamma |k| k$ instead of $\omega = c_0 k - \beta k^3$, the Benjamin–Ono equation is derived instead of the KdV equation as an approximate equation. For example, in the phenomenon of interfacial wave in a two-layer fluid system in which two fluids with different densities overlap, this Benjamin–Ono type dispersion relation may appear.[9]

5.5 WHITHAM EQUATION AND WAVE BREAKING

There are two types of "limit of solution" for gravity waves. One relates to the Stokes wave, i.e., the periodic steady traveling wave solution considering nonlinear effects which we discussed in Chapter 3. In the Stokes wave, the crest of the wave gets more peaky as the wave height increases, and when it reaches a limiting wave height, the crest loses smoothness and reaches an angle of 120°. Let's call this phenomenon "peaking." The critical wave height for deep water gravity waves is known to be about 0.142 times the wavelength. There is no periodic steady traveling wave solution with a wave height exceeding this limiting wave height as discussed in Section 3.4.

Another "limit of solution" relates to unsteady propagation of long waves. As we saw in Chapter 1, when only the effect of nonlinearity is considered as in (5.1), the higher part of the waveform propagates faster. As a result, the front of the waveform becomes steeper as the wave propagates, eventually leading to the appearance of infinitely large surface slope, which we call here "breaking." When water waves hit a structure such as a breakwater, it is known that the magnitude of the impact pressure applied varies greatly depending on whether or not the waves are broken. Thus, the wave breaking is an important problem in engineering.

Although both "peaking" and "breaking" are interesting nonlinear phenomena that actually occur in real water waves, unfortunately neither of them can be reproduced by the KdV equation. For example, in the KdV equation, there is a very stable pulse-like steady traveling wave solution called soliton, but there is no limit to its wave height, and no matter how large the wave height is, the peak of the crest remains smooth and does not become an angle. Also

---

[9]As seawater near the surface of the ocean is warmed and lightened by strong solar radiation, it becomes less likely to mix with cooler and heavier water below it. As a result, in the ocean, a layer where the density of seawater changes rapidly may be formed around several hundred meters of water depth, and it is called "thermocline" or "pycnocline." The same kind of abrupt change in density is often seen also in lakes. Such a situation can be treated approximately as a two-layer fluid system.

for non-stationary evolution of waveform starting from an arbitrary initial waveform, even if the waveform leans forward to some extent due to nonlinearity, the dispersion term (u_{xxx} term) always holds back, and the wave never breaks.

From such a background, Whitham was searching for a simple model wave equation that can reproduce both the peaking and breaking of waves. In the process, he proposed the following model equation:

$$u_t + \alpha u u_x + \int_{-\infty}^{\infty} K(x - \xi) u_\xi(\xi, t)\, d\xi = 0, \qquad (5.51)$$

with a nonlinear term of the same type as KdV equation and a linear integral term. (See, for example, Section 13.14 of [20].) If $u(x, t) = a e^{i(kx - \omega t)}$ is substituted into (5.51) after linearization by ignoring the nonlinear term,

$$- i \omega a e^{i(kx - \omega t)} + \int_{-\infty}^{\infty} K(x - \xi) a i k e^{i(k\xi - \omega t)}\, d\xi = 0,$$

$$\longrightarrow \quad c(k) = \frac{\omega}{k} = \int_{-\infty}^{\infty} K(x - \xi) e^{-ik(x - \xi)}\, d\xi = \int_{-\infty}^{\infty} K(x) e^{-ikx}\, dx. \qquad (5.52)$$

Therefore, by adopting the Fourier transform of the phase velocity $c(k)$ as $K(x)$, it is possible to give (5.51) a desired linear dispersion relation $\omega = k c(k)$.[10]

Among various possibilities, Whitham especially focused on the case of $K(x) = K_w(x) = B e^{-b|x|}$ ($b > 0$). In the following, the equation

$$u_t + \alpha u u_x + \int_{-\infty}^{\infty} B e^{-b|x - \xi|} u_\xi(\xi, t)\, d\xi = 0 \qquad (5.53)$$

that adopts this $K_w(x)$ in (5.51) is called the **Whitham equation**. The linear dispersion relation of (5.53) is given by

$$c(k) = \int_{-\infty}^{\infty} K_w(x) e^{-ikx}\, dx = \frac{2Bb}{k^2 + b^2} \quad \longrightarrow \quad \omega(k) = \frac{2Bbk}{k^2 + b^2}. \qquad (5.54)$$

Whitham has shown that the peaking occurs for (5.53), that is, the steady traveling wave (periodic wave and solitary wave) has an upper limit of the wave height, and the waveform has a sharp angle at the wave crest. For example, when $\alpha = 1$, $B = \pi/4$, $b = \pi/2$, the limiting wave height of steady traveling solitary wave solution is $\frac{8}{9}$, and its waveform and speed are given by

$$u(x, t) = \frac{8}{9} e^{-\frac{\pi}{4}|X|}, \quad X = x - Ut, \quad U = \frac{4}{3}. \qquad (5.55)$$

The wave crest angle of this waveform is 110°. This value is very close to the wave crest angle 120° at the limiting wave height of the real surface gravity wave, but this is merely a coincidence. In

---

[10]Both $c(k)$ and $K(x)$ are assumed to be real even functions.

addition, for the Whitham equation (5.53), it is shown mathematically and numerically that the wave breaking can also occur and the slope of the waveform may diverge [5] and [17]. Figure 5.9 is an example of results of numerical simulations of (5.53). It can be confirmed that the initial waveform $u_0(x) = \sin x$ gradually inclines forward and the front face of the waveform becomes almost vertical in a finite time.

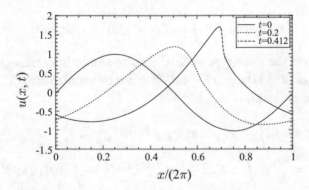

Figure 5.9: A numerical result showing wave breaking in Whitham equation ($B = 2.5, b = 0.5$).

COFFEE BREAK

The solitary wave of the KdV equation was first observed from a scientist's point of view one day in August 1834, at the Union Canal outside Edinburgh, Scotland. John Scott Russell (1808–1882), a shipyard engineer and a researcher on fluid mechanics, was walking along the canal on horseback. At that time the canal was an important route of goods transport. The method of transporting goods was to float a flat-bottomed boat called "barge" on the canal, load the goods on it, and pull it by the horse from the road along the canal. George Stephenson developed a practical steam locomotive from the late 1810s to the 1820s, and Russell seemed to be looking around the canal, examining the possibility of introducing a steamship into canal transport.

At that time, when a certain barge which had been pulled by a horse stopped suddenly, some amount of water had swelled at the bow of the barge, and a beautifully shaped swell started to propagate along the canal at a constant speed with almost no attenuation. Scott Russell was very much interested in the phenomenon, got on the horse and chased the wave for a couple of kilometers, but eventually he lost sight of the wave at a corner of the canal. After that, he conducted research on this beautiful "The great wave of translation" by using an actual canal and a wave flume built in his own garden, and found out various properties of solitary waves and reported them as a paper.

However, G. B. Airy (1801–1892), who later served as the director of the famous Greenwich Observatory for 45 years and had an influence on the scientific community at that time, and Stokes, to whom we mentioned in Chapter 3 as the first person who had introduced nonlinear corrections for periodic surface waves now called the Stokes wave, seem to have been negative that no such solitary wave could be present that would translate with a constant waveform as reported by Russell [4, 10]. The solitary wave of Russell finally recovered its honor when Korteweg and de Vries derived in 1895 an approximate equation for a long wave that later called the KdV equation, and clearly showed that the equation has a solitary wave solution as a steady traveling wave solution, more than 60 years since Russell's "Discovery" of the solitary wave.

Figure 5.10 shows a picture of researchers enjoying generating a solitary waves that Russell might have seen by stopping the boat suddenly near the Union Canal when an international conference focused on solitary waves (solitons) was held in the United Kingdom.

Figure 5.10: Reproduction of Russell's solitary wave in the Union Canal.

5.6   REFERENCES

[1] M. J. Ablowitz and H. Segur. *Solitons and the Inverse Scattering Transform*. SIAM, 1981. DOI: 10.1137/1.9781611970883 91

[2] T. B. Benjamin. Internal waves of permanent form in fluids of great depth. *Journal of Fluid Mechanics*, 29:559–592, 1967. DOI: 10.1017/s002211206700103x 97

[3] T. B. Benjamin, J. L. Bona, and J. J. Mahony. Model equations for long waves in nonlinear dispersive systems. *Philosophical Transactions of the Royal Society*, A227:47–78, 1972. DOI: 10.1098/rsta.1972.0032 96

[4] A. T. Filippov. *The Versatile Soliton*. Birkhäuser, 2000. DOI: 10.1007/978-0-8176-4974-6 101

[5] B. Fornberg and G. B. Whitham. A numerical and theoretical study of certain nonlinear wave phenomena. *Philosophical Transactions of the Royal Society of London*, pp. 373–404, 1978. DOI: 10.1098/rsta.1978.0064 100

[6] C. S. Gardner, J. M. Greene, M. D. Kruskal, and R. M. Miura. Method for solving the Korteweg-de Vries equation. *Physical Review Letters*, pp. 1095–1097, 1967. DOI: 10.1103/physrevlett.19.1095 87

[7] C. S. Gardner, J. M. Greene, M. D. Kruskal, and R. M. Miura. Korteweg-de Vries equation and generalizations. VI. methods for exact solution. *Communications on Pure and Applied Mathematics*, 27:97–133, 1974. DOI: 10.1002/cpa.3160270108 87

[8] J. L. Hammack and H. Segur. The Korteweg-de Vries equation and water waves. II. comparison with experiments. *Journal of Fluid Mechanics*, 65:289–314, 1974. DOI: 10.1017/s002211207400139x 94

[9] R. Hirota. Direct method of finding exact solutions of nonlinear evolution equations. In R. M. Miura, Ed., *Backlund Transformations, the Inverse Scattering Method, Solitons, and their Applications*, number 515 in Lecture Notes in Mathematics, pp. 40–68, Springer, 1976. DOI: 10.1007/bfb0081162 93

[10] R. S. Johnson. *A Modern Introduction to the Mathematical Theory of Water Waves*. Cambridge U.P., 1997. DOI: 10.1017/cbo9780511624056 101

[11] B. B. Kadomtsev and V. I. Petviashvili. On the stability of solitary waves in weakly dispersing media. *Soviet Physics–Doklady*, 15:539–541, 1970. 97

[12] D. J. Korteweg and G. de Vries. On the change of form of long waves advancing in a rectangular canal and on a new type of long stationary waves. *Philosophical Magazine*, 39:422–443, 1895. DOI: 10.1080/14786449508620739 81

[13] P. D. Lax. Integrals of nonlinear equations of evolution and solitary waves. *Communications on Pure and Applied Mathematics*, XXI:467–490, 1968. DOI: 10.1002/cpa.3160210503 92

[14] T. Maxworthy and L. G. Redekopp. A solitary wave theory of the great red spot and other observed features in the Jovian atmosphere. *Icarus*, 29:261–271, 1976. DOI: 10.1016/0019-1035(76)90054-3 95

[15] R. M. Miura, C. S. Gardner, and M. D. Kruskal. Korteweg-de Vries equation and generalizations. II. existence of conservation laws and constants of motion. *Journal of Mathematical and Physics*, 9, 1968. DOI: 10.1063/1.1664701 87

[16] H. Ono. Algebraic solitary waves in stratified fluids. *Journal of the Physical Society of Japan*, 39:1082–1091, 1975. DOI: 10.1143/jpsj.39.1082 97

[17] R. L. Seliger. A note on the breaking of waves. *Proc. of the Royal Society*, A303:493–496, 1968. DOI: 10.1098/rspa.1968.0063 100

[18] M. Shahrill, M. S. F. Chong, and H. N. H. M. Nor. Applying explicit schemes to the Korteweg-de Vries equation. *Modern Applied Science*, 9:200–224, 2015. DOI: 10.5539/mas.v9n4p200 93

[19] T. Taniuti and K. Nishihara. *Nonlinear Waves*. Pitman, 1983. 91

[20] G. B. Whitham. *Linear and Nonlinear Waves*. John Wiley & Sons, 1974. DOI: 10.1002/9781118032954 99

[21] N. J. Zabusky and M. D. Kruskal. Interactions of "solitons" in a collisionless plasma and the recurrence of initial states. *Physical Review Letters*, 15:240–243, 1965. DOI: 10.1103/physrevlett.15.240 85, 86, 87

CHAPTER 6

Modulation and Self-Interaction of a Wavetrain

Since dispersive waves like water wave have different propagation speeds depending on the wavelength, they are automatically sorted according to the wavelength as they travel for a long distance, and reach a state where it looks like a sinusoidal wavetrain with almost uniform wavelength when viewed locally. Such a wavetrain whose wavelength and amplitude change only slowly in space and time is called a "modulated wavetrain" or "quasi-monochromatic wavetrain." In this chapter, we will study the effects of dispersion and nonlinearity on such modulated wavetrains, and the various phenomena resulting from the competition between the two effects.

6.1 MODULATED OR QUASI-MONOCHROMATIC WAVETRAIN

As discussed in Section 3.2, the gravity wave on water surface is dispersive, and the wave propagation velocity depends on the wavelength and frequency. This dispersion becomes weaker as the wavelength λ becomes longer compared to the water depth h, and all waves with wavelengths far longer than h propagate at the same speed of $\sqrt{gh}$. In Chapter 5, we considered those waves whose wavelengths are somewhat shorter than this long wave limit and still had a weak dispersive effect. In the ocean, the area where waves that allow such treatment exist is basically limited to the coastal area. On the other hand, for most of the ocean waves that exist widely on the offshore sea surface, the water depth is much deeper than their wavelengths, mostly corresponding to the deep water wave state, and thus the waves have strong dispersion.

If the wave has strong dispersion, the wave is automatically sorted according to the wavelength and frequency as it propagates. For example, let's assume that a storm develops in a certain sea area and a strong wind blows. Waves of various wavelengths are excited simultaneously by the strong wind. Therefore, the wave field in that area is in a very irregular state including a wide range of wavenumber and frequency components. These waves that are generated by the storm travel through the stormy area to other areas. Then, due to the difference in the propagation velocity caused by the dispersion, the location is gradually divided according to the wavenumber and frequency. For example, let's imagine the situation one day after the waves are generated by the storm. Considering that wave energy travels at the group velocity,[1] a wave with period of 10 s will travel about 670 km from the place where it was generated, while a wave with period of 11 s will travel about 740 km away. Even with these two component waves that differ in period only by 1 s, they will be observed at a distance of 70 km or more from each other one day after departure from the stormy area. As described above, in the dispersive wave, components having different wavelengths and frequencies are spatially separated with time, so when viewed locally, a state in which the wavelength and frequency are constant and close to a uniform wavetrain is realized spontaneously. Such a wavetrain that appears to be a substantially uniform wavetrain locally, but whose amplitude and wavelength are slowly changing is called a **modulated wavetrain**. From the Fourier analysis point of view, the perfectly uniform wavetrain $A \cos(k_0 x - \omega_0 t)$ has a line spectrum in which the energy is concentrated only at wavenumber k_0. On the other hand, the modulated wavetrain has a narrow spectrum centered on a certain wavenumber, and is also called a **quasi-monochromatic wavetrain**.

The fact that the amplitude and wavelength of the wavetrain vary very slowly in the physical space and the fact that the spectrum is very narrow in the spectral space such as frequency space and wavenumber space are equivalent. If we are more conscious of the former aspect, we will call it a "modulated wavetrain," while if we are more conscious of the latter aspect, we will call it a "quasi-monochromatic wavetrain."

6.2 GROUP VELOCITY

6.2.1 GROUP VELOCITY AS PROPAGATION VELOCITY OF MODULATION

Consider a quasi-monochromatic wavetrain whose spectrum is narrow and its amplitude and wavenumber change slowly with respect to x and t. So far, when words such as "long," "fast," etc. are used, we have repeatedly stated that "compare to what" must be clearly understood. For example, in the case of the KdV equation in the previous chapter, the target wave was a "long wave," in the sense that it had a long wavelength λ compared to the water depth h. Then, if δ is introduced as $\delta^2 = h/\lambda$, δ is a small non-dimensional parameter, and by using the perturbation method based on δ, a complex system of basic equations for water waves could be reduced to the much simpler KdV equation. At the same time, it was also possible to understand the range of validity of the KdV equation by evaluating the magnitude of the terms neglected in the process of derivation.

Also in the case of quasi-monochromatic wavetrains to be dealt with in this chapter, when we say "the amplitude or wavenumber is slowly changing," it is necessary to clearly specify "how slow and compared to what." However, let us turn it to the next section, and we will understand it vaguely here to the extent that they hardly change in the length of several wavelengths or

---

[1]We will deal with this subject in detail in the next section.

so. Also, in this section, the effect of nonlinearity will not be considered, so the amplitude is assumed to be infinitesimal.

Since such a quasi-monochromatic wavetrains appear to be approximately sinusoidal locally, local wavenumber k and frequency ω can be introduced. However, these are not constant, but change slowly temporally and spatially. Focusing on a short section Δx in the wavetrain, consider the inflow and outflow of waves during a short time Δt (see Fig. 6.1).[2] Let the waves be traveling from left to right. During Δt, $\omega_A \Delta t/(2\pi)$ waves enter from the left end A of the section, and $\omega_B \Delta t/2\pi$ waves exit from the right end B. On the other hand, the number of waves in this section is given by $k\Delta x/(2\pi)$. In a slowly changing wavetrain where a smooth k and ω can be defined, it is possible to track each of the waves, and the waves do not suddenly appear or disappear. Therefore, the increase or decrease of the number of waves in the interval AB is directly linked to the difference between the inflow and the outflow. From the requirement that the increase of the number of waves in the section per unit time is equal to the net inflow, we obtain

$$\frac{\partial}{\partial t}\left(\frac{k\Delta x}{2\pi}\right) = (\omega_A - \omega_B)\frac{1}{2\pi} = -\frac{\partial \omega}{\partial x}\Delta x\frac{1}{2\pi} \quad \longrightarrow \quad \frac{\partial k(x,t)}{\partial t} + \frac{\partial \omega(x,t)}{\partial x} = 0. \qquad (6.1)$$

This is an equation that may be called the **conservation law of waves**.

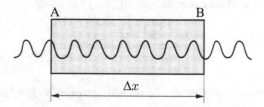

Figure 6.1: Conservation of waves.

Since we are considering wavetrains whose wavelength and frequency change very slowly, it can be thought that the dispersion relation $\omega = \omega(k)$ for the uniform wavetrain holds as it is between the local k and ω in a good approximation. Therefore, we obtain

$$\frac{\partial k}{\partial t} + v_g \frac{\partial k}{\partial x} = 0, \qquad v_g(k) \equiv \frac{d\omega(k)}{dk} \qquad (6.2)$$

immediately from (6.1). This indicates that a constant value of k propagates at a speed of $v_g(k)$. If ω is a function of only k, such as $\omega = \omega(k)$, $k = $ constant immediately means that $\omega = $ constant. Therefore, for ω, the equation of the same form holds, i.e.,

$$\frac{\partial \omega}{\partial t} + v_g \frac{\partial \omega}{\partial x} = 0. \qquad (6.3)$$

[2]Since we are going to investigate changes in k and ω, when we say "short section Δx" and "short time Δt," they are short in the sense that k and ω hardly change, and they are much longer than the wavelength and the period, respectively. Therefore, Δx contains a considerable number of waves, and a considerable number of waves pass through during Δt.

The velocity v_g defined by $v_g = d\omega(k)/dk$ that appears here is called the **group velocity**. Equations (6.2) and (6.3) show that the change of k and ω of the modulated wavetrain are transmitted at the group velocity v_g.

. . . .

[Supplement]

It is only when the property of the medium through which the wave travels are spatially homogeneous and temporally stationary that the dispersion relation is given by $\omega = \omega(k)$, that is, ω is a function of k only. If the medium changes slowly in time, the dispersion relation becomes a form that explicitly include t as in $\omega = \omega(k, t)$, and if the medium changes slowly in space, it has a form that explicitly includes x such as $\omega = \omega(k, x)$.

For example, if the ocean waves travel from offshore to near shore through a region where water depth $h(x)$ gradually decreases, the dispersion relation is given by $\omega = \sqrt{gk \tanh[kh(x)]}$ and depends explicitly on x. In the case of such a dispersion relation of the type $\omega = \omega(k, x)$, a change of ω which occurs due to explicit dependence on x is newly generated, in addition to the usual part brought about through the change of k. In this case, the term $\frac{\partial \omega}{\partial x}$ in (6.1) becomes

$$\frac{\partial \omega}{\partial x} = \frac{\partial \omega}{\partial k}\bigg|_x \frac{\partial k}{\partial x} + \frac{\partial \omega}{\partial x}\bigg|_k, \qquad (6.4)$$

where $\dfrac{\partial \omega}{\partial k}\bigg|_x$ is the partial derivative of ω with respect to k when x is fixed, while $\dfrac{\partial \omega}{\partial x}\bigg|_k$ is the partial derivative of ω with respect to x when k is fixed. Then, if the group velocity v_g is defined as $v_g = \dfrac{\partial \omega}{\partial k}\bigg|_x$, (6.1) gives

$$\frac{\partial k}{\partial t} + v_g \frac{\partial k}{\partial x} = -\frac{\partial \omega}{\partial x}\bigg|_k \ (\neq 0), \qquad (6.5)$$

and k is not constant from the viewpoint of the observer moving at v_g.

However, by considering

$$\frac{\partial \omega}{\partial t} = \frac{\partial \omega}{\partial k}\bigg|_x \frac{\partial k}{\partial t} + \frac{\partial \omega}{\partial x}\bigg|_k \frac{\partial x}{\partial t} = v_g \frac{\partial k}{\partial t}, \qquad (6.6)$$

and multiplying v_g to both sides of (6.1), we see that (6.3) holds as it is even in this nonuniform case. Thus, a constant value of ω still propagates at the group velocity v_g (not constant, though) even if the medium is not homogeneous, so long as it is stationary in time. From this fact, when a swell of a certain frequency travels from offshore to the shore by passing through a region where the water depth h gradually decreases, although the wavelength of the swell changes with place, but the frequency tends to remain constant, and as a result, the wave period takes an equal value no matter where it is measured from the offshore to a shallower place near the shore.[3] Conversely, if the medium is spatially uniform but changes slowly in time, and the dispersion relation is of the form $\omega = \omega(k,t)$, ω is not conserved along with the wave propagation but k is conserved.

In the following, we assume that the medium is spatially uniform and temporally stationary.

. . . .

6.2.2   GROUP VELOCITY AS PROPAGATION VELOCITY OF ENERGY

As mentioned above, the group velocity v_g is defined by $v_g = d\omega(k)/dk$. On the other hand, the velocity c defined by $c = \omega/k$ represents the propagation velocity of the waveform and is called the **phase velocity**. From the relation,

$$v_g = \frac{d\omega}{dk} = \frac{d(ck)}{dk} = c + k\frac{dc}{dk},$$

(6.7)

in dispersive waves (i.e., $dc/dk \neq 0$), $v_g \neq c$ always. For example, in the case of deep water gravity waves whose dispersion relation is given by $\omega = \sqrt{gk}$, $v_g = c/2$.

In the case of a linear wave traveling through a homogeneous stationary medium, there is no exchange of energy between component waves with different wavenumber. Therefore, the fact that the propagation velocity of k is $v_g(k)$ means that the energy of the component wave is also carried at velocity $v_g(k)$. That is, the group velocity is not only the velocity at which the wavenumber and frequency are transmitted, but also the velocity at which the energy is transmitted. This is not a story limited to water surface waves, but applies to any wave phenomena as well. In particular, for gravity waves, in Section 3.3 we calculated the energy density and energy flux of a wave based on the linear sinusoidal wave solution, and confirmed that the energy propagation velocity U defined as the ratio of the energy flux to the energy density certainly agrees with the group velocity $d\omega/dk$.

In the first place, waves can exist and affect other objects only when the energy is transmitted. In this sense, the group velocity, which is the velocity of energy propagation, has more importance than the phase velocity, which is the velocity of waveform propagation. From (6.2), an observer moving at a constant velocity V always sees k such that $V = v_g(k)$ in front of him/her. If there is no energy exchange between different wavenumber components, the energy spectrum does not change with time, and the amount of energy between wavenumber k_1 and k_2 (the shaded part in Fig. 6.2a) does not change with time. On the other hand, in the

[3] This property has already been mentioned in Section 3.2.5 and is used in Example 3.

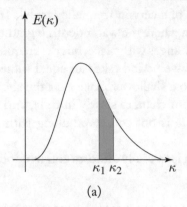
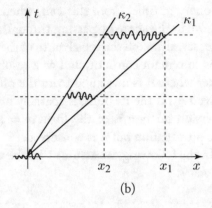

(a)   (b)

Figure 6.2: Energy spectrum and wave propagation in physical space: (a) energy spectrum and (b) the domain in the physical space where wave with k such that $k_1 < k < k_2$ is present.

physical space, after time t, the waves of wavenumber k_1 and k_2 propagate to the position of $x_1 = v_g(k_1)t$ and $x_2 = v_g(k_2)t$, respectively, hence the distance between the two increases in proportion to t (see Fig. 6.2b). The energy of the wave between x_1 and x_2 in the physical space is the energy contained between k_1 and k_2 of the energy spectrum. Since this fixed amount of energy is distributed to the section $x_1 < x < x_2$ which becomes longer in proportion to t, the amplitude of the wave there, which is proportional to the square root of the energy density, decays in time as $1/\sqrt{t}$.

The fact that energy propagates at the group velocity can also be understood from the following consideration. As we saw in Section 3.3, energy is proportional to the square of the amplitude within the linear theory, so the larger the amplitude, the larger the energy should be. For example, when a group of waves are transmitted as shown in Fig. 6.3, the place where the group of waves is located is obviously the place where energy is, so if you look at the moving of the group of waves, it is possible to know the speed of energy propagation immediately.

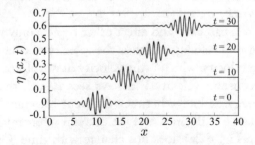

Figure 6.3: Propagation of a wave group.

However, it is not possible to determine the propagation velocity of the energy associated with a wave with a wavenumber k from the motion of such an amplitude pattern. This is because a monochromatic wavetrain consisting of only a certain wavenumber k is a uniform wavetrain of constant amplitude, and even if energy is flowing along the wavetrain, it cannot be seen as a movement of the amplitude pattern. Therefore, in order to discuss the energy propagation in a uniform wavetrain, it is necessary to take a procedure as follows: we once introduce a non-uniformity of the amplitude into the uniform wavetrain in some way, and establish the energy propagation velocity as the propagation speed of the amplitude pattern, then reduce the artificially introduced non-uniformity in order to return to the original uniform wavetrain.

More specifically, this procedure becomes as follows. For a uniform wavetrain $\eta = A \cos(k_0 x - \omega_0 t)$ with wavenumber k_0 and frequency $\omega_0 = \omega(k_0)$, consider a nonuniform wavetrain consisting of superposition of two waves with slightly different wavenumber and frequency

$$\eta = \frac{1}{2}A \cos[(k_0 + \Delta k)x - (\omega_0 + \Delta\omega)t] + \frac{1}{2}A \cos[(k_0 - \Delta k)x - (\omega_0 - \Delta\omega)t], \qquad (6.8)$$

where Δk is a very small wavenumber difference, and $\Delta\omega$ is the resultant small frequency difference. In the limit $\Delta k \to 0$, (6.8) reduces to the original uniform wavetrain $\eta = A \cos(k_0 x - \omega_0 t)$. If we apply the trigonometric addition theorem, we can write (6.8) as

$$\eta = A \cos(\Delta k\, x - \Delta\omega\, t) \cos(k_0 x - \omega_0 t), \qquad (6.9)$$

and the waveform it represents is as shown in Fig. 6.4.

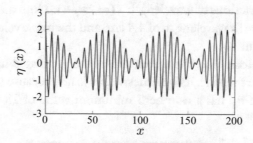

Figure 6.4: Waveform composed of two wavetrains with close wavenumbers ($A = 2$, $k_0 = 1$, $\Delta k = 0.05$).

Of the two cosines of (6.9), the latter $\cos(k_0 x - \omega_0 t)$ represents a uniform wavetrain with wavenumber k_0 and frequency ω_0. On the other hand, the wavenumber and frequency of the former part $A \cos(\Delta k\, x - \Delta\omega\, t)$ are Δk and $\Delta\omega$, respectively. Since $\Delta k \ll k_0$, $\Delta\omega \ll \omega_0$, the change in time and space of this part is very slow compared to the cos behind, and plays a role of specifying the local amplitude of $\cos(k_0 x - \omega_0 t)$. This "amplitude part" travels at a speed of $\Delta\omega/\Delta k$. The propagation of the amplitude pattern is nothing but the propagation of energy, so

$\Delta\omega/\Delta k$ is the propagation velocity of energy. Here, taking the limit of $\Delta k \to 0$ to return to the uniform wavetrain, from the definition of derivative, we obtain $\lim\limits_{\Delta k \to 0} \dfrac{\Delta\omega}{\Delta k} = \dfrac{d\omega(k_0)}{dk}$. Thus, it can be understood that, in the uniform wavetrain of wavenumber k, the energy certainly moves at the group velocity, even though it cannot be seen as a movement of the amplitude pattern.

The far field of linear dispersive waves, that is, the behavior after long-time and long-distance propagation, can be analyzed more precisely by using Fourier analysis and a method of asymptotic evaluation of integrals called the "method of stationary phase." It is somewhat difficult for the level of this book, so I will not touch it here. Readers who are interested should refer to, for example, [11].

6.2.3   EVIDENCES OF ENERGY PROPAGATION AT GROUP VELOCITY

Let us show some evidences from familiar water wave phenomena that show that energy is transmitted at the group velocity.

Waves Caused by a Stone

As shown in Section 3.3, the linear dispersion relation of water waves is given by

$$\omega^2 = gk + \frac{\tau k^3}{\rho}, \tag{6.10}$$

considering both gravity and surface tension as restoring forces. Here, the water depth is assumed to be sufficiently deep compared to the wavelength. Figure 6.5 shows the phase velocity c and the group velocity v_g as functions of wavelength $\lambda(= 2\pi/k)$. Note that the group velocity has a minimum value of 18 cm/s at a wavelength of 4.4 cm, and the phase velocity c for this wavelength is 28 cm/s and is larger than v_g.

When throwing a stone into the water, waves are generated and expand circularly, but after a while the central part regains its quietness. This is not because the waves there have been dissipated out but because v_g has a non-zero minimum value of 18 cm/s, and energy cannot

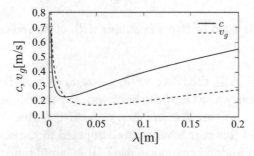

Figure 6.5: The phase and group velocities of capillary gravity waves.

stay near the center. If there were no minimum value of v_g and energy can be transmitted very slowly without limit, the water surface near the center would keep rippling forever. A wave with wavelength of about 4–5 cm, which corresponds to the minimum value of v_g, should be observed just outside the circular area where this quietness is restored. Also, if you look closely, these waves appear from the inner quiescent area as if emerging from nothing. This is a consequence of the fact that $v_g < c$ for these waves.

Difference in the Number of Waves Between Temporal and Spatial Waveforms

Suppose that a group of waves is generated by moving the wave maker installed at one end of a long wave flume 10 times at a frequency ω as shown in Fig. 3.1. If we take a picture of the instantaneous waveform (spatial waveform) of the water surface from the side of the flume, a wave group consisting of 10 waves of wavelength λ corresponding to ω through the dispersion relation will be photographed. Assuming that a wave gauge is installed at a fixed point A downstream of the flume and that the temporal change (temporal waveform) of the height of the free surface is recorded there. When the wave group generated by the wavemaker pass through point A, how many waves will be recorded in the temporal waveform observed there? Since the wave group consists of 10 waves, it is natural to expect that the temporal waveform contains 10 oscillations, but in fact about twice the oscillations are recorded.

The wave group on the left side ($x \approx 10$ m) of Fig. 6.6a shows the initial waveform given at $t = 0$, while the wave group on the right side ($x \approx 28$ m) of the figure shows the spatial waveform after 30 s obtained numerically based on the linearized system of equations of water waves (3.28). The wavelength of the waves that make up the wave group is 1 m, and the group velocity for this wavelength is $v_g = 0.62$ m/s. Although a slight deformation is seen, the wave group can be confirmed to propagate about 18 m in 30 s as expected from the value of v_g.

On the other hand, Fig. 6.6b shows the temporal waveform obtained at point $x = 20$ m for the same case, but the number of oscillations visible in the temporal waveform is clearly larger than the number of waves included in the spatial waveform, and it seems that there are about twice as many. Why does this happen? This seemingly strange phenomenon is also caused by the fact that the waveform is transmitted at phase speed c, but the energy and a wave group are transmitted at group velocity v_g.

Suppose that the wave maker is oscillated n times to generate a wave group consisting of n waves. If the wavelength of the waves that make up the wave group is λ, the spatial length L of the wave group is given by $L = n\lambda$. Since the propagation velocity of a wave group is v_g, the time τ it takes for this wave group to pass through a spatially fixed point A is given by $\tau = L/v_g = n\lambda/v_g$. On the other hand, the wave period T corresponding to the wavelength λ is given by $T = \lambda/c$, where c is the phase velocity corresponding to λ. In the temporal waveform, one oscillation is observed every time T elapses, so the number N of waves observed at the fixed point A is

$$N = \frac{\tau}{T} = \frac{n\lambda}{v_g} \bigg/ \frac{\lambda}{c} = n \times \frac{c}{v_g}. \tag{6.11}$$

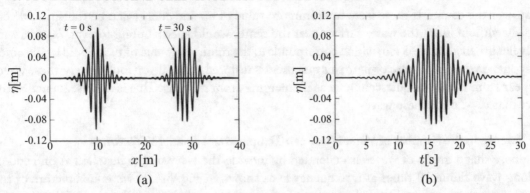

Figure 6.6: Difference in number of waves in the spatial waveform (a) and the temporal waveform (b) ($\lambda = 1$ m, $v_g = 0.62$ m/s).

Therefore, the number N of waves in the temporal waveform is c/v_g times the number n of waves in the spatial waveform. In the case of deep water gravity waves whose dispersion relation is given by $\omega = \sqrt{gk}$, $c/v_g = 2$, so the temporal waveform will contain twice as many waves as the spatial waveform.[4]

Figure 6.7 shows the same numerical results as Fig. 6.6 in a different way. The horizontal axis is x and the vertical axis is t. The value of t increases downward. The value of $\eta(x, t)$ at each (x, t) is shown in colors. Red corresponds to the place where η is largest, that is , the crest of the waves, and blue corresponds to the place where η is minimum, that is, the trough of the waves. In the green background corresponding to $\eta = 0$, it can be seen that the wave group moves about 20 m from $x = 10$ to $x = 30$ in about 32 s. However, it can be seen that the red and blue portions corresponding to the crests and troughs in the wave group move along lines closer to

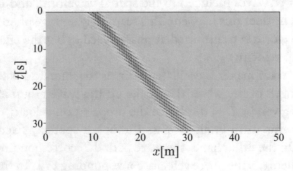

Figure 6.7: Propagation of a wave group ($\lambda = 1$ m, $v_g = 0.62$ m/s).

---

[4]In the case of the capillary waves whose dispersion relation is $\omega \propto k^{3/2}$, $c/v_g = 2/3$, and conversely, the number of waves in the temporal waveform is smaller than the number of waves in the spatial waveform.

the horizontal compared with the lines going down to the right corresponding to the movement of the wave group as a whole. "Closer to the horizontal" means that the spatial location changes more in a same interval of time, that is, the speed is faster. If we look at the figure closely, we can see that the waves emerge from the rear end of the wave group, and they moves in the wave group at a speed (= phase speed) faster than the speed of the wave group (= group velocity), and that they disappear at the front end of the group. The fact that the waves move at a higher speed in the wave group than the wave group itself is the cause of the increase in the number of waves in the temporal waveform.

Identifying the Storm that Caused the Swell

Imagine that the waves produced by a distant storm are transmitted across the ocean and are arriving at the shore where you are now. A storm produces waves of various wavelengths simultaneously, but when the waves leave the area of the storm and propagate as swell, swells with longer wavelengths (i.e., lower frequency) propagate faster according to the dispersion relation $\omega = \sqrt{gk}$ of gravity wave in deep water. Therefore, even if the swells are generated simultaneously by a same storm, the swell with a lower frequency reaches a distant coast faster. Suppose that a swell with a period of 23 s had arrived four days ago on the beach where you are, but then the period gradually became shorter, and today, a swell of period of 10 s is arriving. The group velocity of the wave with a period of 23 s is about 18 m/s (about 1,500 km/day), and the group velocity of the wave with a period of 10 s is about 7.8 m/s (about 670 km/day). As shown in Fig. 6.8, these waves are traced back to the past at each group velocity, and by looking at the point where the two intersect, it can be guessed that these waves were generated by a storm which occurred 7 days before about 5,000 km from this coast. The direction of the storm can also be inferred by observing the direction in which the swell comes from offshore where the waves are not yet affected by refraction due to the bottom topography. If the distance, direction, and the date have been estimated in this way, it is possible to identify the storm that caused the swells that have been arriving at the coast for the last several days.

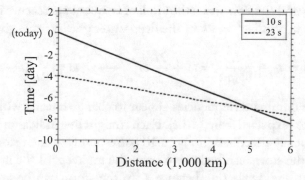

Figure 6.8: Identification of the source of swells from change of period.

Figure 6.9: Observation points used in Snodgrass et al. (1966) [10].

Using this principle, Snodgrass et al. (1966) [10] conducted a large-scale study to find out where the long waves with periods of 10–20 s in the summertime Hawaii and California, which are ideal for surfing, are coming from. Figure 6.9 shows the observation site they have deployed which literally spans the North and South Pacific Ocean from Antarctic to Alaska. Figure 6.10 is an example of the frequency spectrum that they observed at an observation point. The horizontal axis is the date, and the vertical axis is the frequency in mHz. The intensity distribution is indicated by contour lines as well as the grayscale. The darker parts represent the part with larger spectral intensity. For example, focusing on the part surrounded by the dotted line in the figure, on the morning of August 6, the wave of 45 mHz in frequency, that is a period of 22 s, was prominent, but the frequency rose gradually with time, and it can be seen that on the night of August 8, about 75 mHz, that is, swell with a period of 13 s became dominant. If the time when the storm produced these swells is t_0, the distance between the storm and the observation point is L, and the time when the swell of frequency ω reaches the observation point is $t(\omega)$, the dispersion relation $\omega^2 = gk$ of the deep water gravity wave gives

$$t(\omega) = t_0 + \frac{L}{v_g(\omega)} = t_0 + \frac{2L}{g}\omega \quad \longrightarrow \quad \omega = \frac{g}{2L}(t - t_0). \tag{6.12}$$

In Fig. 6.10, the dominant frequencies appear to change linearly with time, which is consistent with the behavior expected from (6.12). Each straight line indicating the temporal change of the dominant frequency described in Fig. 6.10 corresponds to each storm. From (6.12), the occurrence time t_0 of the storm can be known from the intercept of the horizontal t axis of the corresponding straight line, while the distance L to the storm can be known from the slope ($= g/2L$) of the corresponding straight line. The smaller the slope of the straight line, the farther the storm is from the observation point. For example, we can see that the series of swells

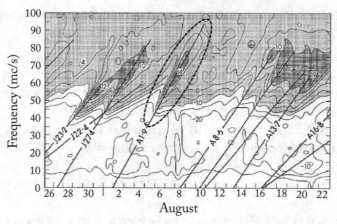

Figure 6.10: Example of frequency spectrum obtained by Snodgrass et al. (1966) [10].

surrounded by dotted line in Fig. 6.10 were produced by a storm about 6,400 km away on August 1st.

Analysis of these observations collected by Snodgrass et al. revealed that the majority of the swells reaching Hawaii and California in summer come from around the Antarctic continent far beyond the equator. In the Southern Hemisphere, there is a western wind that winds around the earth equivalent to the westerly wind in the Northern Hemisphere. But, since the Southern Hemisphere has less land area, the wind is less likely to be decelerated than the Northern Hemisphere, and the wind is stronger. Especially in winter (summer in the Northen Hemisphere), strong cyclones develop and large storms often blow, and the area around 40° and 50° south are called "roaring forties" and "furious fifties" and are feared by sailors. Although it might be disappointing for Japanese surfers, these swells are blocked by Australia and the islands of Indonesia and do not reach Japan.

6.3   NONLINEAR SCHRÖDINGER EQUATION: EQUATION GOVERNING MODULATION

In the previous section, we showed that, starting from the conservation law of wave (6.1), that the changes in wavenumber, frequency, and amplitude are transmitted at group velocity in a quasi-monochromatic wavetrain. In this section, we will investigate the propagation of modulation in a quasi-monochromatic wavetrain more systematically.

6.3.1   CONTRIBUTION FROM DISPERSION: LINEAR SCHRÖDINGER EQUATION

In order to describe the evolution of a quasi-monochromatic wavetrain, it is convenient to push all deviations from the "carrier wave," which is a complete uniform wavetrain with wavenumber k_0 and the corresponding frequency ω_0, into the "complex amplitude" as follows. First of all, let us ignore nonlinearity and consider only the effect of dispersion.

Suppose that the initial value $\eta(x, 0)$ of the target physical quantity (e.g., free surface displacement) $\eta(x, t)$ is represented by its Fourier transform as

$$\eta(x, 0) = \int_0^\infty f(k) e^{ikx}\, dk + \text{c.c.}. \tag{6.13}$$

Since $\eta(x, t)$ is assumed to be real, it can thus be written as a sum with a complex conjugate. In this case, if the linear dispersion relation is $\omega = \omega(k)$, then the solution $\eta(x, t)$ at any later time t is given by

$$\eta(x, t) = \int_0^\infty f(k) e^{i[kx - \omega(k)t]}\, dk + \text{c.c.}. \tag{6.14}$$

Here, if we introduce the wavenumber deviation $\kappa = k - k_0$ and expand $\omega(k)$ around k_0 in a Taylor series as

$$\omega(k) = \omega(k_0 + \kappa) = \omega(k_0) + \omega'(k_0)\kappa + \frac{1}{2}\omega''(k_0)\kappa^2 + \cdots, \tag{6.15}$$

then (6.14) can be written as[5]

$$\eta(x, t) = e^{i(k_0 x - \omega_0 t)} \int_{-\infty}^\infty f(k_0 + \kappa) e^{i\kappa x} e^{-i[\omega_0'\kappa + (1/2)\omega_0''\kappa^2 + \cdots]t}\, d\kappa + \text{c.c.}. \tag{6.16}$$

If we introduce the **complex amplitude** $a(x, t)$ by

$$a(x, t) \equiv \int_{-\infty}^\infty f(k_0 + \kappa) e^{i\kappa x} e^{-i\{\omega_0'\kappa + (1/2)\omega_0''\kappa^2 + \cdots\}t}\, d\kappa, \tag{6.17}$$

and write a in polar form as $a = |a|e^{i\theta}$, then (6.16) can be written as

$$\eta(x, t) = a(x, t) e^{i(k_0 x - \omega_0 t)} + \text{c.c.} = 2|a|\cos(k_0 x - \omega_0 t + \theta). \tag{6.18}$$

Here, $e^{i(k_0 x - \omega_0 t)}$ represents a uniform wavetrain corresponding to the carrier wave, and all the information on amplitude and wavenumber modulation is included in $a(x, t)$ Thus, a wave field

---

[5]Let me say a few words about making the range of integration of κ to $(-\infty, \infty)$. The object of this section is a quasi-monochromatic wave whose wavenumber spectrum has energy only in a narrow range around k_0. Assuming that the width of the wavenumber spectrum is Δk, $f(k)$ in (6.14) has a nonzero value only in the interval of $k_0 - \Delta k \leq k \leq k_0 + \Delta k$, which corresponds to $-\Delta k \leq \kappa \leq \Delta k$. In the integration with respect to κ in (6.16), $f(k)$ is substantially zero in the part beyond the interval $-\Delta k \leq \kappa \leq \Delta k$, and it does not matter if we extend the range of integration to $-\infty \leq \kappa \leq \infty$ for convenience.

that has a spectrum concentrated in a narrow range around wavenumber k_0 can be treated as a single wavetrain (6.18) by introducing the complex amplitude $a(x, t)$.

From the definition (6.17) of a,

$$\frac{\partial a}{\partial t} = -i \int_{-\infty}^{\infty} \left(\omega_0' \kappa + \frac{1}{2} \omega_0'' \kappa^2 + \cdots \right) f(k_0 + \kappa)\, e^{\{\cdots\}}\, d\kappa, \tag{6.19a}$$

$$\frac{\partial a}{\partial x} = i \int_{-\infty}^{\infty} \kappa f(k_0 + \kappa)\, e^{\{\cdots\}}\, d\kappa, \tag{6.19b}$$

$$\frac{\partial^2 a}{\partial x^2} = -\int_{-\infty}^{\infty} \kappa^2 f(k_0 + \kappa)\, e^{\{\cdots\}}\, d\kappa, \tag{6.19c}$$

then, in an approximation that ignores $O(\kappa^3)$, $a(x, t)$ satisfies

$$i \left\{ \frac{\partial a}{\partial t} + \omega_0' \frac{\partial a}{\partial x} \right\} + \frac{1}{2} \omega_0'' \frac{\partial^2 a}{\partial x^2} = 0. \tag{6.20}$$

This equation is called the **linear Schrödinger equation**. As you can see from the above derivation process, this argument is not limited to water wave problem but holds generally.

In order to understand the meaning of (6.20) more clearly, let us check the magnitude of each term. Let A and Δk be the typical magnitudes of the amplitude a and the spectrum width, respectively. Because we ignore nonlinear effects here, the amplitude must be small enough. However, a itself is an amount with a dimension, and it is meaningless to argue smallness against it.[6] In the case of the deep water gravity waves adopted as an example here, it is the non-dimensional amplitude Ak_0 that must be small as discussed in Section 3.4. It is also important in this section that the spectrum width is narrow, but Δk is also a quantity with a dimension of [1/length], and the smallness of itself is meaningless. A small dimensionless parameter that means "narrow band" is $\Delta k/k_0$, that is, the spectrum width Δk normalized by the carrier wavenumber k_0. For example, we expanded $\omega(k)$ in Taylor series around k_0 and ignored the third and higher order terms in (6.15). For this truncation to be rational, it is naturally required that the second term is much smaller than the first term, and the third term is in turn much smaller than the second term, and so on. If we estimate the magnitudes of κ, ω', ω'' by Δk, ω_0/k_0, ω_0/k_0^2, respectively, the ratio of the second term to the first term, the ratio of the third term to the second term, and the ratio of other subsequent two terms of (6.15) is always $O(\Delta k/k_0)$.

With these things in mind, and from (6.19), the order of magnitude of a_t, a_x, a_{xx} can be estimated as

$$a \sim A, \quad a_t \sim \omega_0 \frac{\Delta k}{k_0} A, \quad a_x \sim \Delta k A, \quad a_{xx} \sim (\Delta k)^2 A. \tag{6.21}$$

---

[6] The magnitude of a quantity with a dimension can vary widely with the unit system. For example, even if the same length of 1 m, it becomes a small value of 0.001 when measured in km, while it becomes a large value of 1000 when measured in mm. It is only for non-dimensional quantities that can rationally argue if it is large or small.

Then the order of magnitude of each term of (6.20) can be estimated as follows:

$$a_t \sim \omega_0 \left(\frac{\Delta k}{k_0} \right) A, \quad \omega_0' a_x \sim \omega_0 \left(\frac{\Delta k}{k_0} \right) A, \quad \omega_0'' a_{xx} \sim \omega_0 \left(\frac{\Delta k}{k_0} \right)^2 A, \qquad (6.22)$$

and we can see that the first two terms have the same order of magnitude, while the third term is relatively smaller by $(\Delta k / k_0)$. Therefore, in the case of extremely narrow band, i.e., $\frac{\Delta k}{k_0} \ll 1$, it is conceivable to ignore the third term and consider only the first two terms. This part means that $a(x, t)$ translates without change of form at the group velocity ω_0' corresponding to the carrier wavenumber k_0, which is exactly what we discussed in Section 6.2.

Various temporal and spatial scales are involved in the description of the modulated wavetrain. First there is a time scale T_0 and a space scale L_0 that characterize the carrier wave, which of course are of the order of the carrier period and the wavelength, so

$$T_0 \sim \frac{1}{\omega_0}, \quad L_0 \sim \frac{1}{k_0}. \qquad (6.23)$$

Next, there is a time scale T_1 and a space scale L_1 of modulation, that is, the scale of time and space with which we can notice that the carrier that looks like a uniform wavetrain when viewed locally with scales T_0 and L_0 is actually modulated slowly. These can be evaluated by a/a_t and a/a_x, respectively. According to the order estimate of a_t, a_x above,

$$T_1 \sim \frac{a}{a_t} \sim \frac{1}{\omega_0} \frac{k_0}{\Delta k} \sim \left(\frac{k_0}{\Delta k} \right) T_0, \quad L_1 \sim \frac{a}{a_x} \sim \frac{1}{k_0} \frac{k_0}{\Delta k} \sim \left(\frac{k_0}{\Delta k} \right) L_0. \qquad (6.24)$$

Therefore, the temporal and spatial lengths T_1, L_1 required to see the modulation of the wavetrain becomes longer than the scales of the carrier T_0, L_0 as $\frac{\Delta k}{k_0}$ becomes smaller. The third term of (6.20) represents the deviation from the pure translation of $a(x, t)$ with carrier group velocity ω_0' represented by the first two terms, that is, it expresses the net deformation of $a(x, t)$. Since this term is smaller by $(\Delta k / k_0)$, the time and space scales T_2, L_2 in which the deformation of the modulation waveform caused by this term becomes prominent become ever longer and are given as

$$T_2 \sim \left(\frac{k_0}{\Delta k} \right)^2 T_0, \quad L_2 \sim \left(\frac{k_0}{\Delta k} \right)^2 L_0. \qquad (6.25)$$

We showed in Fig. 6.6 that the results of numerical simulation of wave group propagation. If we look at the figure carefully, we can notice that after 30 s the shape of the envelope of the wave group is somewhat gentler and wider than the initial one. This kind of deformation is exactly the effect of this third term. As can be seen from the coefficient ω_0'', the third term is a term due to the difference in group velocity between the wavenumbers that compose the narrow spectrum, and hence it is called **group velocity dispersion**.

6.3.2   CONTRIBUTION FROM NONLINEARITY: MODE GENERATION AND RESONANCE

In the discussion so far, we have ignored the nonlinear effect assuming that the amplitude is extremely small, and have investigated only the dispersive effect that arises from having a narrow but finite band width. In the following, conversely, the effect of dispersion is neglected by assuming that the spectrum is a line spectrum with no width, and instead the influence of nonlinearity on the evolution of complex amplitude $a(x, t)$ is considered. First, we will study as preparation two things: "mode generation" by nonlinear interaction and "resonant interaction."

Mode Generation by Nonlinear Interaction

Suppose that $\eta(x, t)$ is composed of two component waves as

$$\eta(x,t) = \left(a_1 e^{i\theta_1} + \text{c.c.}\right) + \left(a_2 e^{i\theta_2} + \text{c.c.}\right), \quad \theta_i = k_i x - \omega_i t \ (i = 1, 2) \tag{6.26}$$

at its lowest-order $O(a)$. If the governing equation of the system or its boundary conditions contains a second-order nonlinear term in η, as it can be seen from

$$\eta^2 = \left\{a_1^2 e^{2i\theta_1} + a_2^2 e^{2i\theta_2} + a_1 a_2 e^{i(\theta_1 + \theta_2)} + a_1 a_2^* e^{i(\theta_1 - \theta_2)}\right\} + \text{c.c.} + 2\left(|a_1|^2 + |a_2|^2\right), \tag{6.27}$$

wave components with new wavenumber and frequency combinations made from the sum or difference of wavenumber and frequency combination (k_1, ω_1), (k_2, ω_2) such as $(2k_1, 2\omega_1)$, $(2k_2, 2\omega_2)$, $(k_1 \pm k_2, \omega_1 \pm \omega_2)$ are automatically generated.[7] Here, a^* represents the complex conjugate of a. If we write this schematically,

$$\eta = \begin{cases} a_1 e^{i\theta_1} + \text{c.c.} \\ a_2 e^{i\theta_2} + \text{c.c.} \end{cases} \implies \eta^2 = \begin{cases} a_1^2 e^{2i\theta_1} + \text{c.c.} \\ a_2^2 e^{2i\theta_2} + \text{c.c.} \\ a_1 a_2 e^{i(\theta_1 + \theta_2)} + \text{c.c.} \\ a_1 a_2^* e^{i(\theta_1 - \theta_2)} + \text{c.c.} \\ |a_1|^2, \ |a_2|^2. \end{cases} \tag{6.28}$$

[7]However, it may not be appropriate to call these oscillating components that are newly generated by nonlinearity "waves" just because they have the form $a \, e^{i(kx - \omega t)}$. Because, as studied so far, in order for the sinusoidal wave $a \, e^{i(kx - \omega t)}$ to be the "true wave" that the system allows, k and ω need to satisfy the dispersion relation required by the governing equations (& boundary condition) of the system. In the case considered here, (k_1, ω_1), (k_2, ω_2) of the lowest order components are originally supposed to satisfy this dispersion relation. But there is no guarantee at all that the combinations $(2k_1, 2\omega_1)$, $(2k_2, 2\omega_2)$, $(k_1 \pm k_2, \omega_1 \pm \omega_2)$ that are generated due to nonlinearity also satisfy the dispersion relation. If the system is truly dispersive (i.e., the dispersion relation is not $\omega = c_0 k$), it is rather exceptional that they satisfy the dispersion relation. This point will be discussed again in the next chapter.

If $\eta(x, t)$ is composed of only a single monochromatic wavetrain in the lowest $O(a)$, then components such as

$$\eta = a\,e^{i\theta} + \text{c.c.}, \tag{6.29a}$$

$$\eta^2 = \left(a^2 e^{2i\theta} + \text{c.c.}\right) + 2|a|^2, \tag{6.29b}$$

$$\eta^3 = \left(a^3 e^{3i\theta} + 3|a|^2 a e^{i\theta}\right) + \text{c.c.} \tag{6.29c}$$

will be generated due to nonlinear interaction with itself. Here, the fact that a component with the same time-space dependence $e^{i\theta}$ as the original wave is excited at the third order $O(a^3)$ has an important implication as discussed below.

Resonant Interaction

When an oscillatory system with a natural frequency ω_0 like a spring or a pendulum is subjected to an oscillatory external force with a frequency ω, its response $y(t)$ is governed by the ordinary differential equation

$$\frac{d^2 y}{dt^2} + \omega_0^2\, y = F \cos \omega t \tag{6.30}$$

under the linear approximation. As we discussed in Section 4.3.2, when $\omega = \omega_0$, resonance occurs. In this case, the general solution of (6.30) is given by

$$y(t) = A \cos(\omega_0 t + \theta) + \frac{F}{2\omega_0}\, t\, \sin \omega_0 t, \tag{6.31}$$

and the amplitude of the inhomogeneous part (second term) increases indefinitely in proportion to time. As an analogy of this phenomenon, resonant interference can also occur if a wavelike external force $f \propto e^{i(kx - \omega t)}$ acts on a wave represented by $a\,e^{i(k_0 x - \omega_0 t)}$. However, unlike oscillations, waves are oscillatory not only in time but also in space, so in order to realize a resonant state, the wavenumber must match as well as the frequency, i.e., $k = k_0$, $\omega = \omega_0$.

If the basic wave field $O(a)$ consists of a monochromatic wave $a\,e^{i\theta} + \text{c.c.}$ ($\theta = k_0 x - \omega_0 t$), the nonlinear interaction with itself produces a component of $|a|^2 a\,e^{i\theta}$ at $O(a^3)$ as (6.29) shows. This component is in a relationship of resonance with the fundamental wave, so it acts as a resonant external force on the fundamental wave. (6.31) indicates that the rate of change da/dt of the amplitude a of an oscillatory system subjected to a resonant external force is proportional to the magnitude F of the external force. By analogy, it can be inferred that a term proportional to $|a|^2 a$ would appear as a nonlinear effect in the equation governing the temporal change of the complex amplitude $a(x, t)$. If the dimensionless parameter representing the smallness of the amplitude is denoted as ϵ, then the magnitude of this term is $O(\epsilon^3)$, so the effect of nonlinear self-interaction is expected to appear prominently on a space-time scale of T_0/ϵ^2 and L_0/ϵ^2.

6.3.3   NONLINEAR SCHRÖDINGER EQUATION

As we saw previously, the deformation of the complex amplitude $a(x, t)$ brought about by the effect of group velocity dispersion becomes noticeable on a time scale of $T_0(k_0/\Delta k)^2$. On the other hand, as seen above, the influence of nonlinearity on $a(x, t)$ appears through a term of the form $|a|^2 a$, and the effect is expected to become prominent on a time scale of $T_0/(Ak_0)^2$. As in the case of the Burgers equation and the KdV equation, when several different effects coexist, interesting and important phenomena appear in the space-time scale where they have the same magnitude and compete with each other. In the present case, this corresponds to the situation where $(\Delta k/k_0) \sim (Ak_0)$, i.e., the small parameter $\Delta k/k_0$ expressing the narrowness of the spectrum (i.e., weakness of group velocity dispersion) and the small parameter Ak_0 expressing the smallness of the amplitude (i.e., weakness of nonlinearity) are equally small. If we derive an equation which governs the time-space evolution of the complex amplitude from the original set of governing equations for various types of wave phenomena in different physical systems such as fluids, plasmas, etc., we generally obtain an equation of the following form, as expected from the above intuitive argument,

$$i \left(a_t + \omega_0' a_x\right) + p \, a_{xx} = q \, |a|^2 a. \tag{6.32}$$

Here p, q are both real numbers, and p is always given by $\omega''(k_0)/2$, as can be seen from the linear Schrödinger equation (6.20). This equation is called the **nonlinear Schrödinger equation**, and was first derived for deep-water surface gravity waves by Zakharov (1968) [13]. In the following, we call it **NLS equation** for short. We also eliminate the second term by introducing a coordinate system moving at a group velocity $\omega'(k_0)$ and represent it as

$$i a_t + p \, a_{xx} = q \, |a|^2 a. \tag{6.33}$$

Of course, the NLS equation should not be derived by intuition as above, but it should be derived more systematically using the perturbation method starting from the basic equations which govern the physical system. However, for the case of waver waves, it requires quite complicated calculations to actually derive the NLS equation from the original set of governing equation. Then in the following we will show specifically the process of derivation of the NLS equation for an imaginary system which is governed by the KdV equation by using the multiple-scale method.

· ·

EXAMPLE 1: DERIVATION OF THE NLS EQUATION BY MULTIPLE SCALE METHOD

In a system governed by the KdV equation[8]

$$u_t + c_0 u_x + \beta u_{xxx} = -\alpha u u_x, \tag{6.34}$$

derive an NLS equation that describes the space-time evolution of the complex amplitude $a(x, t)$ of the quasi-monochromatic wavetrain with carrier wavenumber k_0 and frequency ω_0.

[Answer]

Let ϵ be a small dimensionless parameter that indicates the smallness of the amplitude. As discussed above, it is reasonable to assume that the small quantity representing the narrowness of the spectrum $\Delta k / k_0$ is comparable to the smallness of nonlinearity, and therefore ϵ at the same time represents the smallness of $\Delta k / k_0$. As described above, the description of a quasi-monochromatic wavetrain includes various time and space scales, such as those of the carrier wave itself, those in which the modulation is visible, and those in which the effects of group velocity dispersion and the nonlinear self-interaction are visible, and so on. To express such phenomena with various scales well, it is convenient to use the perturbation method called the multiple scale method introduced in Chapter 4. In this method, multiple time and space variables are introduced as shown below.

Let x_0, t_0 be variables suitable for viewing the carrier wave itself. This is a space-time variables originally present, so

$$x_0 = x, \quad t_0 = t. \tag{6.35}$$

If we look at the modulation of the wavetrain by using x_0, t_0, these variable need to change to a very large value of around $1/\epsilon$. Therefore we introduce new slow variables by

$$x_1 = \epsilon x_0, \quad t_1 = \epsilon t_0. \tag{6.36}$$

Similarly, for the effects of group dispersion and nonlinear self-interaction to become visible, it is necessary for x_0, t_0 to become even larger values of $O(1/\epsilon^2)$. So we also introduce new variables

$$x_2 = \epsilon^2 x_0, \quad t_2 = \epsilon^2 t_0, \tag{6.37}$$

to handle such long distance and time appropriately. If you want to consider higher-order nonlinear effects and dispersion effects, you may introduce variables x_n, t_n ($n \geq 3$) that change more slowly in a similar manner if necessary.

---

[8]As discussed in Chapter 5, the KdV equation is an important nonlinear wave equation that can be derived in many physical systems, taking into account both weak nonlinearity and weak dispersion. However, in this example, we will forget all the physical meaning of the KdV equation and the premise of its derivation, and will use it as a starting point just to show the typical derivation process of the NLS equation.

The dependent variable $u(x, t)$ is approximated by a series expansion in ϵ, and each term is considered to be functions of x_n, t_n introduced above. That is,

$$u(x, t) = \sum_{j=1} \epsilon^j u_j = \epsilon u_1 + \epsilon^2 u_2 + \cdots, \quad u_j = u_j(x_0, t_0, x_1, t_1, \cdots), \quad (j = 1, 2, \ldots).$$

(6.38)

As the independent variables x, t are expanded to several variables, differential operations are also expanded as

$$\frac{\partial}{\partial x} = \frac{\partial}{\partial x_0}\frac{\partial x_0}{\partial x} + \frac{\partial}{\partial x_1}\frac{\partial x_1}{\partial x} + \frac{\partial}{\partial x_2}\frac{\partial x_2}{\partial x} + \cdots = \frac{\partial}{\partial x_0} + \epsilon\frac{\partial}{\partial x_1} + \epsilon^2\frac{\partial}{\partial x_2} + \cdots, \quad (6.39a)$$

$$\frac{\partial}{\partial t} = \frac{\partial}{\partial t_0}\frac{\partial t_0}{\partial t} + \frac{\partial}{\partial t_1}\frac{\partial t_1}{\partial t} + \frac{\partial}{\partial t_2}\frac{\partial t_2}{\partial t} + \cdots = \frac{\partial}{\partial t_0} + \epsilon\frac{\partial}{\partial t_1} + \epsilon^2\frac{\partial}{\partial t_2} + \cdots. \quad (6.39b)$$

Due to this property, this multiple scale method is also called the **derivative expansion method**.

Substituting (6.38) and (6.39) into (6.34), organizing according to the power of ϵ, and solving from the lower order. First, the problem with $O(\epsilon)$ is as follows:

$$L_0[u_1] = 0, \quad L_0 \equiv \frac{\partial}{\partial t_0} + c_0\frac{\partial}{\partial x_0} + \beta\frac{\partial^3}{\partial x_0^3}. \quad (6.40)$$

Since we are discussing here the modulation of a nearly monochromatic wave, the monochromatic carrier wave should be adopted as the solution of the lowest order $O(\epsilon)$, so

$$u_1 = a(x_1, t_1, \ldots)e^{i\theta} + \text{c.c.}, \quad \theta = k_0 x_0 - \omega_0 t_0, \quad D(k_0, \omega_0) = \omega_0 - ck_0 + \beta k_0^3 = 0, \quad (6.41)$$

where k_0 and ω_0 satisfy the dispersion relation $\omega(k) = c_0 k - \beta k_0^3$ required by the operator L_0. $D(k, \omega)$ is a polynomial of k and ω such that the following equation holds when L_0 operates on $e^{i(kx_0 - \omega t_0)}$,

$$L_0\left[e^{i(kx_0 - \omega t_0)}\right] = -iD(k, \omega)\, e^{i(kx_0 - \omega t_0)}. \quad (6.42)$$

Equation (6.40) is nothing but the linearized version of (6.34) except that x and t are written as x_0 and t_0, and (6.41) is its sinusoidal wave solution. However, it should be noted that the amplitude a is allowed to depend on slow variables such as x_1 and t_1. This dependence of a on the slow scales represents the modulation of wavetrain.

The problem of $O(\epsilon^2)$ is given as follows:

$$L_0[u_2] + \left(\frac{\partial}{\partial t_1} + c_0\frac{\partial}{\partial x_1} + 3\beta\frac{\partial^3}{\partial x_0^2 \partial x_1}\right) u_1 = -\alpha u_1\frac{\partial u_1}{\partial x_0}. \quad (6.43)$$

Substituting the solution (6.41) of $O(\epsilon)$ into this yields

$$L_0[u_2] = -\left\{\frac{\partial a}{\partial t_1} + (c_0 - 3\beta k_0^2)\frac{\partial a}{\partial x_1}\right\} e^{i\theta} + \text{c.c.} - i\alpha k_0 a^2 e^{2i\theta} + \text{c.c..} \quad (6.44)$$

Here, $\{\cdots\}\mathrm{e}^{i\theta}$ on the right side is a solution of the corresponding homogeneous problem $L_0[u] = 0$. This is exactly the same situation when a resonance occurs. Therefore, if this term is present, a secular term will appear in u_2 which grows infinitely in proportion to θ, and the perturbation expansion will break down in a short time. Then as a non-secular condition, we obtain

$$\frac{\partial a}{\partial t_1} + \left(c_0 - 3\beta k_0^2\right)\frac{\partial a}{\partial x_1} = 0. \tag{6.45}$$

Note here that the coefficient $(c_0 - 3\beta k_0^2)$ of $\partial a/\partial x_1$ is equal to $\omega'(k_0)$, i.e., the group velocity of the carrier wave. You might feel like we are requesting something self-serving, but this non-secular condition (6.45) automatically gives the correct result that the complex amplitude a propagates at the carrier group velocity. Also, as a result of this, we obtain

$$u_2 = \frac{\alpha k_0}{D(2k_0, 2\omega_0)} a^2 \mathrm{e}^{2i\theta} + \text{c.c.} + \phi(x_1, \ldots, t_1, \ldots) \tag{6.46}$$

as the solution to the problem of (6.43). Here, the part $\mathrm{e}^{2i\theta}$ is a particular solution of the in-homogeneous problem, and ϕ is a "DC component" (real number) equivalent to a constant of integration.

The next $O(\epsilon^3)$ problem becomes as follows:

$$L_0[u_3] = -\left(\frac{\partial}{\partial t_1} + c_0\frac{\partial}{\partial x_1} + 3\beta\frac{\partial^3}{\partial x_0^2 \partial x_1}\right)u_2$$

$$- \left(\frac{\partial}{\partial t_2} + c\frac{\partial}{\partial x_2} + 3\beta\frac{\partial^3}{\partial x_0 \partial x_1^2} + 3\beta\frac{\partial^3}{\partial x_0^2 \partial x_2}\right)u_1$$

$$- \alpha\left(u_1\frac{\partial u_2}{\partial x_0} + u_2\frac{\partial u_1}{\partial x_0} + u_1\frac{\partial u_1}{\partial x_1}\right). \tag{6.47}$$

Substituting (6.41) and (6.46) for u_1 and u_2, respectively, brings about on the right-hand side a term proportional to $\mathrm{e}^{i\theta}$ and a constant term that would cause secular terms in u_3 as in the case of $O(\epsilon^2)$. If we require that the coefficients of these terms be both zero to prevent the breakdown of the perturbation expansion, we obtain

$$\text{const. term:} \quad \frac{\partial \phi}{\partial t_1} + c_0\frac{\partial \phi}{\partial x_1} + \alpha\frac{\partial |a|^2}{\partial x_1} = 0, \tag{6.48a}$$

$$\mathrm{e}^{i\theta} \text{ term :} \quad i\left(\frac{\partial a}{\partial t_2} + \omega_0'\frac{\partial a}{\partial x_2}\right) + \frac{1}{2}\omega_0''\frac{\partial^2 a}{\partial x_1^2} = \left\{\frac{\alpha^2 k_0^2}{D(2k_0, 2\omega_0)}|a|^2 + \alpha k_0\phi\right\}a. \tag{6.48b}$$

As (6.48a) shows, the DC component ϕ is produced by spatial modulation of $|a|$. Since a just translates at the group velocity ω_0' on the (x_1, t_1) space-time scale as shown by (6.45), it is reasonable to assume that ϕ produced by a has the same (x_1, t_1) dependencies as a. Then from (6.48a)

$$\phi = \alpha|a|^2/(\omega_0' - c_0), \tag{6.49}$$

and substituting this into (6.48b) gives

$$i\left(\frac{\partial a}{\partial t_2} + \omega_0'\frac{\partial a}{\partial x_2}\right) + \frac{1}{2}\omega_0''\frac{\partial^2 a}{\partial x_1^2} = \left\{\frac{\alpha^2 k_0^2}{D(2k_0, 2\omega_0)} + \frac{\alpha^2 k_0}{\omega_0' - c_0}\right\}|a|^2 a. \qquad (6.50)$$

By adding ϵ^2 times of (6.45) and ϵ^3 times of (6.50), and using the relationship (6.45) to restore the original variable x and t, and at the same time, replacing ϵa with a in order to let a be the amplitude of u, we finally obtain

$$i\left(\frac{\partial a}{\partial t} + \omega_0'\frac{\partial a}{\partial x}\right) + \frac{1}{2}\omega_0''\frac{\partial^2 a}{\partial x^2} = \left\{\frac{\alpha^2 k_0^2}{D(2k_0, 2\omega_0)} + \frac{\alpha^2 k_0}{\omega_0' - c_0}\right\}|a|^2 a, \qquad (6.51)$$

as a condition that the secular term does not occur up to $O(\epsilon^3)$. This is the NLS equation that describes the space-time evolution of the complex amplitude of a quasi-monochromatic wavetrain in a system governed by the KdV equation (6.34).[9]   ♣

. . . .

As can be seen from the example above, it is not so easy to derive the NLS equation, even if the governing equation is such a simple one like (6.34). Deriving the NLS equation for the surface gravity waves starting from the far more complex basic equations (3.28) of water waves requires considerable mathematical ability and patience. This task has been completed by Hasimoto and Ono (1972) [5]. After tedious calculations, the equation they finally obtained is the NLS equation of the form (6.33) itself, and its coefficients p and q are given by

$$p = \frac{1}{2}\omega_0'' = -\frac{g}{8k_0\sigma\omega_0}\left[\{\sigma - k_0 h(1-\sigma^2)\}^2 + 4k_0^2 h^2\sigma^2(1-\sigma^2)\right], \qquad (6.52a)$$

$$q = \frac{g^2 k_0^4}{2\omega_0^3}\left[\frac{1}{\omega_0' - gh}\{4c_0^2 + 4(1-\sigma^2)c_0\omega_0' + gh(1-\sigma^2)^2\} + \frac{1}{2\sigma^2}(9 - 10\sigma^2 + 9\sigma^4)\right], \qquad (6.52b)$$

where h is the water depth, k_0 is the carrier wavenumber, ω_0 is the carrier frequency determined from k_0 by the linear dispersion relation $\omega = \sqrt{gk\tanh(kh)}$, c_0 $(= \omega_0/k_0)$ is the phase velocity of the carrier, ω_0' is the group velocity of the carrier, and $\sigma = \tanh(k_0 h)$. The coefficient p is always negative for any combination of k_0 and h, while the coefficient q changes its sign, and is positive when $k_0 h > 1.363$ and negative when $k_0 h < 1.363$. This change of sign at some specific value of $k_0 h$ has very important implications for the stability of the wavetrain, as shown in Section 6.4.

[9]According to (6.51), the coefficient of the nonlinear term diverges when $D(2k_0, 2\omega_0) = 0$ and $\omega_0' - c_0 = 0$, and this analysis breaks down. In fact, the former corresponds to the situation called "harmonic resonance" and the latter corresponds to the situation called "long wave short wave resonance." We will treat these subjects in Section 7.4 of the next chapter.

6.3.4   ENVELOPE SOLITON SOLUTION

Various analytical solutions are known for the NLS equation (6.33), and the most famous of them is the **envelope soliton** solution

$$a(x,t) = a_0 \text{sech} \left\{ \sqrt{\frac{-q}{2p}} \, a_0 x \right\} \exp\left(-\frac{i}{2} q a_0^2 t \right). \quad (a_0 \text{: arbitrary real number}). \qquad (6.53)$$

This represents a pulse-like wave group as shown in Fig. 6.11a. The fine waves inside represents the carrier wave, and the solution $a(x,t)$ of the NLS equation defines the shape of its envelope.

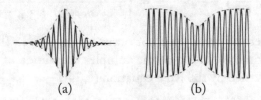

(a)   (b)

Figure 6.11: Envelope soliton (a) and envelope hole (b).

Like the soliton solution (5.12) of the KdV equation, the envelope soliton of the NLS equation is narrower as the amplitude is larger. However, in the KdV soliton, the width δ is about $1/\sqrt{a}$, whereas the width of the envelope soliton of the NLS is about $1/a$. KdV soliton was realized on the balance between the nonlinear term uu_x and the dispersion term u_{xxx}. When the representative amplitude and width of the wave are a and δ, respectively, the order of these terms are estimated as $uu_x \sim O(a^2/\delta)$ and $u_{xxx} \sim O(a/\delta^3)$, and hence it is required that $\delta \sim 1/\sqrt{a}$. On the other hand, the envelope soliton of the NLS equation is built on the balance of the group velocity dispersion a_{xx} and the nonlinearity $|a|^2 a$. The magnitudes of these terms are estimated as $O(a/\delta^2)$ and $O(a^3)$, respectively, so it is necessary that $\delta \sim 1/a$ for them to balance with each other. As the expression in its square root shows, the envelope soliton solution (6.53) exists only when the coefficients p and q have opposite sign ($pq < 0$).[10]

The NLS equation, like the KdV equation, is shown to be one of the so-called "soliton equations" that can be solved by the inverse scattering method [14]. Therefore, the envelope soliton of the NLS is also a "soliton" as its name suggests, and it has been confirmed in actual water tank experiments that it exhibits particle-like stability against interactions [12].

. . . .

EXAMPLE 2: CHECKING THE ENVELOPE SOLITON SOLUTION

Confirm that the envelope soliton solution (6.53) satisfies the NLS equation (6.33).

[10]Conversely, in the case of $pq > 0$, there exists a solution in which an isolated depression is transmitted in the uniform wavetrain as that shown in Fig. 6.11b, which cannot exist when $pq < 0$. This is called the **envelop hole** or **dark soliton**. In this respect, the envelope soliton is also called the **bright soliton**.

[Answer]

First, let $y(x) = \text{sech}\, x \,(= 1/\cosh x)$. Then, from

$$(\text{sech}\, x)' = -\tanh x\, \text{sech}\, x, \quad (\text{sech}\, x)'' = \text{sech}\, x - 2\text{sech}^3 x, \tag{6.54}$$

$y(x)$ is known to satisfy the differential equation

$$y'' = y - 2y^3. \tag{6.55}$$

If we write the envelope soliton (6.53) as

$$a(x,t) = a_0 \,\text{sech}\, \xi\, E, \quad \xi = \sqrt{\frac{-q}{2p}}\, a_0 x, \quad E = \exp\left(-\frac{i}{2} q a_0^2 t\right), \tag{6.56}$$

then

$$ia_t = i a_0 \,\text{sech}\, \xi\, E \left(-\frac{i}{2} q a_0^2\right) = \frac{1}{2} q a_0^3 \,\text{sech}\, \xi\, E, \tag{6.57a}$$

and

$$pa_{xx} = pa_0 \frac{d^2 \text{sech}\xi}{d\xi^2} \left(\frac{d\xi}{dx}\right)^2 E = pa_0 (\text{sech}\, \xi - 2\text{sech}^3 \xi) \left(\frac{-q}{2p} a_0^2\right) E$$

$$= -\frac{1}{2} q a_0^3 (\text{sech}\xi - 2\text{sech}^3 \xi)\, E. \tag{6.57b}$$

Then,

$$ia_t + pa_{xx} = q a_0^3 \text{sech}^3 \xi\, E = q|a|^2 a, \tag{6.58}$$

and the NLS equation certainly holds.

♣

. . . .

The NLS equation is invariant to the Galilean transform

$$\tilde{x} = x - Vt, \quad \tilde{t} = t, \quad \tilde{a} = a \exp\left[-i\frac{V}{2p}x + i\frac{V^2}{4p}t\right], \tag{6.59}$$

and

$$ia_t + pa_{xx} = q|a|^2 a \quad \longleftrightarrow \quad i\tilde{a}_{\tilde{t}} + p\tilde{a}_{\tilde{x}\tilde{x}} = q|\tilde{a}|^2 \tilde{a} \tag{6.60}$$

holds. This invariance physically corresponds to the indeterminacy of the carrier wavenumber k_0. The wavetrain we are dealing with has a spectral width of about $\Delta k \sim \epsilon k_0$, and it is our disposal to choose which wavenumber is the carrier wavenumber. The seemingly complex phase factor $e^{i(\cdots)}$ in (6.59) corresponds to replacing the current carrier wavenumber k_0 with a new carrier wavenumber $\tilde{k}_0$ for which the group velocity differs by V (i.e., $\tilde{k}_0 - k_0 = V/2p$). Using (6.59), we can easily obtain an envelope soliton propagating at an arbitrary velocity V from a stationary envelope soliton (6.53). However, remembering the physical background of the NLS equation and the assumptions made in the process of derivation, V must be a small amount of $O(\epsilon)$, however arbitrary it may be. It should be noted that the NLS equation (6.33) is also invariant under the transformation

$$t \to \lambda^2 t, \quad x \to \lambda x, \quad a \to \lambda^{-1} a \quad (\lambda \neq 0).$$ (6.61)

. . . .

EXAMPLE 3: CHECKING GALILEAN TRANSFORMATION INVARIANCE

Confirm that, if $a(x, t)$ is a solution of the NLS equation (6.33), then $\tilde{a}(x, t)$ given by

$$\tilde{a}(x, t) = a(x - Vt, t) \exp\left[i\frac{V}{2p}x - i\frac{V^2}{4p}t\right]$$ (6.62)

is also a solution of (6.33).

[Answer]
    Let

$$\tilde{E} \equiv \exp\left[i\frac{V}{2p}x - i\frac{V^2}{4p}t\right].$$ (6.63)

Then

$$\tilde{a}_t = \left[a_x(-V) + a_t - i\frac{V^2}{4p}a\right]\tilde{E},$$ (6.64a)

$$\tilde{a}_x = \left[a_x + i\frac{V}{2p}a\right]\tilde{E},$$ (6.64b)

and

$$\tilde{a}_{xx} = \left[a_{xx} + i\frac{V}{2p}a_x + \left(a_x + i\frac{V}{2p}a\right)\left(i\frac{V}{2p}\right)\right]\tilde{E},$$ (6.64c)

and from these it can be seen that

$$i\tilde{a}_t + p\tilde{a}_{xx} = \cdots = (ia_t + pa_{xx})\,\tilde{E} = q|a|^2a\,\tilde{E} = q|\tilde{a}|^2\tilde{a}.$$ (6.65)

♣

. . . .

Similar to the KdV soliton, the width of the envelope soliton of the NLS is determined by the amplitude as (6.53) indicates. However, unlike the KdV soliton which becomes faster as the amplitude increases, there is no special relationship between the propagation velocity and the amplitude of the envelope soliton. This independence between amplitude and velocity can also be seen from the above Galilean invariance.[11] Reflecting this, the NLS equation allows a new type of solution in which multiple solitons with different amplitudes propagate at the same speed without separation, which can be called a "bound state of solitons." Such solutions are called **breathers** because they propagate with periodic oscillation (see, for example, [9]).

As a special case of the breather solution, the following solution called "Peregrine breather" is especially well known [8],

$$u(x,t) = a_0 \exp\left(-i q a_0^2 t\right) \left[-1 + \frac{4(1 - 2i q a_0^2 t)}{1 - \frac{2q}{p} a_0^2 x^2 + 4 q^2 a_0^4 t^2}\right]. \tag{6.66}$$

This solution approaches a uniform wavetrain in both the limit $x \to \pm\infty$ and the limit $t \to \pm\infty$, and a wave group appears around $(x, t) = (0, 0)$ only once which has three times the amplitude of the uniform wavetrain. In recent years, a phenomenon in which an unusually high wave suddenly appears as if it has come from nowhere in a relatively calm ocean has attracted attention, and it is called **freak waves** or **rogue waves**. Some researchers consider the breather solution of the NLS equation to be a simple and effective model for such phenomena of rogue waves (see, for example, [2] and [3]).

When you drain the water of the bathtub, you can see the water flows out with the form of a thin vortex. In the fluid dynamics, the limiting state where the vorticity is concentrated within a thin string with zero thickness is called a "vortex filament." Intuitively, it is a pattern of motion that has nothing to do with wave phenomena, but it has been shown that the motion of this vortex filament can also be reduced to the NLS equation under certain approximation [4]. Therefore, there is a pattern of motion, sometimes called "Hasimoto soliton," in which a twist of the vortex filament propagates stably along the filament which corresponds to the envelope soliton solution of the NLS equation.

6.4   MODULATIONAL INSTABILITY

We learned in the previous section that the modulation of a quasi-monochromatic wavetrain is governed by the nonlinear Schrödinger equation. Using this fact, we can discuss the stability of a uniform wavetrain to modulation as shown next.

---

[11]The NLS equation can be solved by the inverse scattering method. In other words, there exits a linear eigenvalue problem such that the solution $a(x, t)$ of the NLS equation is included in the coefficient and that its discrete eigenvalues remain constant when $a(x, t)$ evolves according to the NLS equation. And as with the KdV equation, one envelope soliton is associated with one discrete eigenvalue of the problem, and this invariance of discrete eigenvalue is the origin of outstanding stability and self-holding ability of envelope soliton of the NLS equation. However, unlike the KdV equation, the linear eigenvalue problem corresponding to the NLS equation is not self-adjoint, and the eigenvalues generally take complex numbers. According to the inverse scattering method for the NLS equation, the real part of each discrete eigenvalue specifies the speed of the corresponding soliton, while the imaginary part specifies its amplitude. For more detail, refer to [14].

6.4.1   STOKES WAVES AND ITS STABILITY

As mentioned in Section 3.4, G. G. Stokes first introduced nonlinear effects to surface gravity waves about the mid 19th century. He derived the following approximate solution, now called the Stokes wave, for the surface displacement $\eta(x, t)$ of a wavetrain propagating at a constant velocity without changing its shape in water of infinite depth:

$$\eta(x, t) = A \cos \theta + \frac{1}{2} A^2 k \cos 2\theta + O(A^3), \quad \theta \equiv kx - \omega t, \quad \omega = \sqrt{gk} \left(1 + \frac{1}{2} A^2 k^2\right).$$
(6.67)

A particularly important point of this solution is that the frequency ω depends not only on the wavenumber k but also on the amplitude A, and waves of larger amplitudes propagate faster even though the wavelength is the same. Since Stokes, research on Stokes waves has been actively carried out, for example, to carry out the Stokes-like amplitude expansion to higher orders to obtain approximate solutions with wider effective range, or to mathematically discuss the convergence of the Stokes expansion itself, etc.

However, whether it is a solution that satisfies the governing equation for water waves and whether it can actually be observed on the ocean surface or in a wave tank experiment are completely different stories. The property of **stability** is related there. For example, let us think about laying a hemispherical bowl with its face down and putting a small ball at its top as shown in Fig. 6.12a. The top of the bowl corresponds to the extremum (local maximum) of the potential energy, and the situation where the ball rests there is an equilibrium that satisfies the balance of forces. However, the ball will fall off the hemisphere with a slightest deviation from the top, and it is almost impossible to actually realize such a situation. On the other hand, the bottom of the bowl facing upward as shown in Fig. 6.12b is also an extremum (local minimum) of the potential energy, so the balance of forces is established there and the ball can stay there. In this case, even if the ball slightly deviates from the bottom for some reason, a force automatically works to return the ball to the bottom, and the ball does not deviate greatly from the point of balance. As in the former case, if the deviation increases rapidly when a small deviation occurs for some reason, the equilibrium is said to be **unstable**, while if the deviation does not increase in time as in the latter case, it is said to be **stable**. There is no difference that they are both solutions of the equation of force balance, but the former with no stability is not actually realized.

(a)   (b)

Figure 6.12: The concept of stable and unstable situations: (a) unstable and (b) stable.

It has been known since Stokes that the Stokes wave, that is, a periodic wavetrain considering nonlinearity exits as an approximate solution that accurately satisfies the basic equations of water waves. However, no attention seems to have been given to its stability for a long time. However, finally in 1967, over 100 years since Stokes, Benjamin, and Feir for the first time showed both theoretically and experimentally that the Stokes wave is unstable against some kind of small perturbation, and it is impossible to generate a beautifully uniform wavetrain corresponding to the Stokes wave in a wave flume no matter how hard one tries [1].

6.4.2   STABILITY ANALYSIS BASED ON NLS EQUATION

The analysis by Benjamin and Feir is based on in-depth physical insights into the mechanisms that causes instability, but the method of analysis seems somewhat less clear from today's point of view. As shown below, the same results as them can be derived more easily by starting from the NLS equation that they did not know at that time. First, note that the solution of (6.33) which does not depend on x, that is,

$$a(x,t) = a_0\, e^{-iqa_0^2 t}, \tag{6.68}$$

corresponds to the Stokes wave (6.67). Here, a_0 can always be made a real positive number by shifting the origin of t, so we will assume it that way. According to the relation (6.18), $\eta(x,t)$ corresponding to $a(x,t)$ of (6.68) is given by

$$\eta(x,t) = a(x,t)e^{i(k_0 x - \omega_0 t)} + \text{c.c.} = 2a_0 \cos\left[k_0 x - (\omega_0 + qa_0^2)t\right]. \tag{6.69}$$

This represents a uniform wavetrain whose frequency is modified from the value ω_0 of the linear theory by an amount qa_0^2 proportional to the square of the amplitude.

The coefficients p and q of the NLS equation for surface gravity waves shown in (6.52) becomes

$$p = -\frac{\omega_0}{8k_0^2}, \quad q = 2k_0^2 \omega_0, \tag{6.70}$$

at the limit of infinite depth $k_0 h \to \infty$, where $\omega_0 = \sqrt{gk_0}$. If this value of q is used, the frequency ω of the wavetrain of (6.69) becomes $\omega = \omega_0\left[1 + \frac{1}{2}(2a_0)^2 k_0^2\right]$, which matches the frequency of the Stokes wave.[12] Thus, it can be seen that the uniform wavetrain solution (6.68) of the NLS equation corresponds to the Stokes wave. Therefore, by examining the stability of the uniform wavetrain solution (6.68) based on the NLS equation, we can know the stability of the Stokes wave, at least against long wavelength perturbations that can be treated as modulation.

Suppose that the amplitude and the phase of the uniform wavetrain solution are slightly perturbed like

$$a = (a_0 + \tilde{a}) \exp\left[-iqa_0^2 t + i\tilde{\theta}\right], \tag{6.71}$$

[12]Although the second harmonic component of the Stokes wave cannot be known from the NLS equation itself, it is obtained at the $O(\epsilon^2)$ stage of the derivation process of the NLS equation, and it agrees with the second harmonic component of the Stokes expansion (6.67), too.

where $\tilde{a}$ and $\tilde{\theta}$ are both small real numbers. Substituting this into the NLS equation, and retaining only the first-order terms with respect to $\tilde{a}$ and $\tilde{\theta}$, we obtain the following system of equations that governs the disturbances:

$$\tilde{a}_t + pa_0\tilde{\theta}_{xx} = 0, \quad \tilde{\theta}_t - p\tilde{a}_{xx}/A_0 + 2qa_0\tilde{a} = 0. \tag{6.72}$$

Since this is a system of linear partial differential equations with constant coefficients, any solution can be represented by superposition of the fundamental mode

$$\begin{pmatrix} \tilde{a} \\ \tilde{\theta} \end{pmatrix} = \begin{pmatrix} \hat{a} \\ \hat{\theta} \end{pmatrix} e^{i(Kx-\Omega t)} + \text{c.c.} \tag{6.73}$$

If this fundamental mode is stable for all wavenumber K, an arbitrary small perturbation superimposed on the uniform wavetrain will not grow, so the uniform wavetrain is stable. On the other hand, if there is at least one fundamental mode that increases with time, the disturbance with the wavenumber K will spontaneously grow, so the uniform wavetrain is unstable.

Substituting (6.73) into (6.72), we obtain the "dispersion relation"

$$\Omega^2 = p^2 K^2 \left(K^2 + 2qa_0^2/p\right), \tag{6.74}$$

between K and Ω. According to this, if $pq > 0$, real K always gives real Ω, so that all the fundamental modes are neutrally stable. In this case, even if a small perturbation is added to the uniform wavetrain, it does not grow temporally, so the uniform wavetrain solution (6.68), that is, the Stokes wave is stable. On the other hand, if $pq < 0$, then $\Omega^2 < 0$ and Ω is a pure imaginary number for K satisfying

$$|K| < \sqrt{-2q/p}\, a_0. \tag{6.75}$$

According to (6.73), this means that the fundamental mode of such a K grows exponentially in time with the growth rate of Im Ω, and hence the uniform wavetrain is unstable. The growth rate Im Ω takes the maximum value $|qa_0^2|$ for $K = \sqrt{-q/p}\, a_0$. As (6.75) shows, this instability occurs only for those modes whose wavenumber K is as small as the amplitude a_0. Returning to the original $\eta(x, t)$, this means that only the wavenumber components near the carrier wavenumber k_0 are unstable and grow, and as a result it promotes the modulation of the wavetrain. From this property, this instability is called the **sideband instability**, **modulational instability**, or **Benjamin–Feir instability** from the name of its discoverers.

Figure 6.13 shows the growth rate as a function of wavenumber K, in which the growth rate (vertical axis) and the wavenumber (horizontal axis) are normalized with the maximum growth rate $|qa_0^2|$ and the corresponding wavenumber $\sqrt{-q/p}\, a_0$, respectively.

In the case of deep-water gravity waves analyzed by Benjamin and Feir, $p = -\omega_0/8k_0^2 < 0$ and $q = 2\omega_0 k_0^2 > 0$, hence $pq < 0$, and the uniform wavetrain is unstable. Substituting these expressions of p, q into (6.74), we can immediately obtain the growth rate equivalent to the

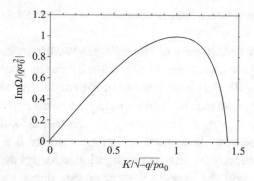

Figure 6.13: Growth rate $\mathrm{Im}\,\Omega$ of modulational instability as a function of modulation wavenumber K.

results of Benjamin and Feir. Then, the condition of modulational instability corresponding to (6.75) becomes

$$\frac{|k - k_0|}{k_0} < 2\sqrt{2}\,A_0 k_0, \quad (A_0 = 2a_0 \text{ is the amplitude of } \eta) \tag{6.76}$$

which indicates that only a narrow range of k around the carrier wavenumber k_0 has a nonzero growth rate. When the water depth h is finite, p and q are complicated functions of $k_0 h$ as shown in (6.52). From these expressions, it can be seen that p is always negative for all $k_0 h$, but q changes sign as $q > 0$ when $k_0 h > 1.363$ and $q < 0$ when $k_0 h < 1.363$. This implies that the gravity wave becomes unstable to modulation when the water depth becomes deeper than about 1/5 of the wavelength, and it can no longer propagate as a beautifully uniform wavetrain.

The NLS equation (6.33) is a very general wave equation that governs nonlinear modulated wavetrains in various media not limited to water waves. In addition, the modulational instability is an instability that occurs whenever the coefficient p of the nonlinear term $|a|^2 a$ and the coefficient q of the group velocity dispersion term a_{xx} have opposite signs, so it is an important phenomenon commonly observed in wave phenomena in wide range of research fields.

6.4.3 INTUITIVE UNDERSTANDING OF MODULATIONAL INSTABILITY

The occurrence of modulational instability when $pq < 0$ can be intuitively understood as follows. Let $p < 0$ and $q > 0$. Since $p = \frac{1}{2}\omega_0'' = \frac{1}{2}dv_g(k_0)/dk$, $p < 0$ means that the group velocity v_g, which is the velocity of energy propagation, is a decreasing function of k around k_0. On the other hand, q gives the coefficient of the frequency correction $q|a_0|^2$ due to nonlinearity as can be seen from the relation between the uniform solution (6.68) of the NLS equation and the corresponding actual waveform $\eta(x, t)$ of the water surface given by (6.69). From this, $q > 0$

means that ω of the wave, and hence the phase velocity, is an increasing function of the amplitude around k_0.

Suppose that only a certain part of the uniform wavetrain has a slightly larger amplitude than the surrounding area, as shown in Fig. 6.14. Since the phase velocity is an increasing function of amplitude ($q > 0$), the phase velocity of the carrier wave is a little faster in the part where the amplitude is large, so that the wave spacing becomes narrower and the wavenumber increases in the front of that part, and conversely, on the rear side, the wave spacing increases and the wavenumber decreases. Then, since the group velocity is a decreasing function of k (since $p < 0$), the group velocity (= energy propagation velocity) decreases at the front of the part where k is increased, and the forward flow of energy decreases. On the other hand, the group velocity increases with the decrease of k on the rear side of the part, and the inflow of energy from the rear increases. Thus, if the amplitude of a part of the wavetrain is increased for some reason, the net energy inflow (= inflow minus outflow) toward the part will occur automatically. As a result, the amplitude of this part will be further increased, and the initial slight nonuniformity of amplitude will be further enhanced. As shown in Example 4 below, the same intuitive understanding is possible when nonuniformity occurs in the wavenumber instead of the amplitude.

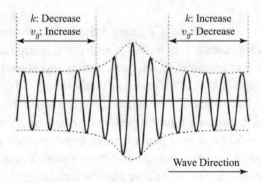

Figure 6.14: Intuitive explanation of modulational instability (in the case of amplitude nonuniformity).

EXAMPLE 4: INTUITIVE UNDERSTANDING OF MODULATIONAL INSTABILITY IN THE CASE OF WAVENUMBER NONUNIFORMITY

Let $p < 0$, $q > 0$ as above. When the wavenumber k becomes larger than the surroundings at a certain part of the uniform wavetrain as shown in Fig. 6.15, explain intuitively that the nonuniformity will increase.

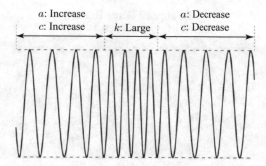

Figure 6.15: Intuitive explanation of modulational instability (in the case of wavenumber nonuniformity).

[Answer]

Since the group velocity is a decreasing function of k from $p < 0$, the group velocity (= energy velocity) decreases and the energy flux decreases in the central part where k is larger than in the surrounding area. As a result, the energy flow to the front part decreases, and the amplitude decreases there. Conversely, due to the slow down of energy flow at the central part, there occurs an accumulation of energy in the rear part, and the amplitude increases there. Then, since the phase velocity is an increasing function of the amplitude (since $q > 0$), in the front part where the amplitude decreases, the phase velocity of the carrier wave decreases, and hence the outflow of the carrier wave is suppressed. Conversely, in the rear part where the amplitude increases, the phase velocity of the carrier wave increases, and hence the inflow of the carrier wave is enhanced. As a result, if there is a part where k is larger than the surrounding area, more and more carrier waves will be concentrated there, and the nonuniformity of k will be further enhanced. ♣

....

6.4.4   MODULATIONAL INSTABILITY AND FREAK WAVE

As mentioned in Section 6.3.4, in recent years, ocean wave phenomenon called "freak waves" or "rogue waves" has attracted attention and has been studied extensively. Freak waves are large waves that suddenly appear in relatively calm ocean wave fields. According to [7], at least 22 large cargo ships sank and 525 people died in the Pacific and Atlantic Oceans between 1969 and 1994 due to encounters with freak waves. Figure 6.16 shows one of the best-known examples of freak waves called "New Year's wave." It can be seen that one large wave suddenly appears that exceeds 25 m in wave height among the waves with an average wave height of about 10 m. This wave height record was observed on the platform of an offshore oil field called Draupner (Fig. 6.17) in the North Sea. The name of the wave has come from the fact that it was observed on January 1, 1995.

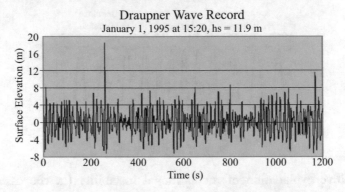

Figure 6.16: Record of surface displacement of the "New Year's Wave" (from [6]).

Figure 6.17: Draupner offshore oil field platform (from [6], photographed by Øyvind Hagen).

During tracing the time evolution of water waves by numerical simulation of the system of basic equations (3.28), we sometimes observe in a computer a single large wave, which should be called a freak wave, appears as shown in Fig. 6.18.

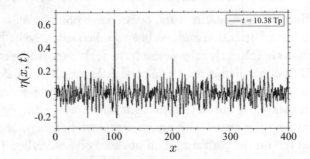

Figure 6.18: A typical freak wave appearing in numerical simulation of water wave.

Figure 6.19 is an example of numerical simulation result to reproduce the modulational instability. The system of basic equations of gravity waves, i.e., (3.28) with $\tau = 0$, is faithfully solved using the same program as that used to obtain the result shown in Fig. 6.18. The only difference between the two cases is in the choice of the initial conditions. In the case of Fig. 6.18, an irregular waveform suitable for ocean waves is initially adopted, while in the case of Fig. 6.19, we adopted a sinusoidal carrier wave of $Ak_0 = 0.07$ and superimposed on it a small modulation of wavelength of 7 times longer than the carrier (dotted line in the figure). Since the initial amplitude of the modulation disturbance is only 1/50 of the amplitude of the carrier wave, the waveform at $t = 0$ looks almost uniform. The small disturbance (sideband components), which satisfies the condition (6.75) and hence is unstable, gradually grows with time, and after 200 periods, the waveform like shown by the solid line in the figure appears. It can be seen that, as a result of the nonuniformity in the wave height brought about by the modulational instability, large water surface displacements more than twice that of the initial waveform occur spontaneously even though there is no extra energy supply from outside of the system. From these facts, many researchers think that modulational instability is one of the key mechanisms for understanding freak wave phenomena.

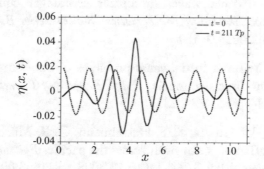

Figure 6.19: Appearance of a large wave due to modulational instability.

The research on freak waves originally started with the phenomenon of ocean waves, but in recent years, both theoretical and experimental researches on freak waves are actively progressing even in completely different physical systems, such as optical fibers, quantum mechanical system called "Bose-Einstein condensates," etc.

6.5   REFERENCES

[1] B. Benjamin and J. E. Feir. The disintegration of wave trains on deep water. part 1. theory. *Journal of Fluid Mechanics*, 27:417–430, 1967. DOI: 10.1017/s002211206700045x 133

[2] A. Chabchoub. Tracking breather dynamics in irregular sea state conditions. *Physical Review Letters*, 117:144103, 2016. DOI: 10.1103/physrevlett.117.144103 131

[3] K. B. Dysthe and K. Trulsen. Note on breather type solutions of the NLS as models for freak waves. *Physica Scripta*, T82:48–52, 1999. DOI: 10.1238/physica.topical.082a00048 131

[4] H. Hasimoto. A soliton on a vortex filament. *Journal of Fluid Mechanics*, 51:477–485, 1972. DOI: 10.1017/s0022112072002307 131

[5] H. Hasimoto and H. Ono. Nonlinear modulation of gravity waves. *Journal of the Physical Society of Japan*, 33:805–811, 1972. DOI: 10.1143/jpsj.33.805 127

[6] S. K. Haver. A possible freak wave event measured at the draupner jacket January 1, 1995, 2004. http://www.ifremer.fr/web-com/stw2004/rogue/fullpapers/walk_on_haver.pdf 138

[7] C. Kharif and E. Pelinovsky. Physical mechanisms of the rogue wave phenomenon. *European Journal of Mechanics B/Fluid*, 22:603–634, 2003. DOI: 10.1016/j.euromechflu.2003.09.002 137

[8] D. H. Peregrine. Water waves, nonlinear scrödinger equations and their solutions. *Journal of the Australian Mathematical Society Series B*, 25:16–43, 1983. DOI: 10.1017/s0334270000003891 131

[9] J. Satsuma and N. Yajima. Initial value problems of one-dimensional self-modulation of nonlinear waves in dispersive media. *Suppl. Progress of Theoretical Physics*, 55:284–306, 1974. DOI: 10.1143/ptps.55.284 131

[10] F. E. Snodgrass, G. W. Groves, K. F. Hasselmann, G. R. Miller, W. H. Munk, and W. H. Powers. Propagation of ocean swell across the pacific. *Philosophical Transactions on the Royal Society of London*, A259, 1966. DOI: 10.1098/rsta.1966.0022 116, 117

[11] G. B. Whitham. *Linear and Nonlinear Waves*. John Wiley & Sons, 1974. DOI: 10.1002/9781118032954 112

[12] H. C. Yuen and B. M. Lake. Nonlinear dynamics of deep-water gravity waves. *Advances in Applied Mechanics*, 22:67–229, 1982. DOI: 10.1016/s0065-2156(08)70066-8 128

[13] V. E. Zakharov. Stability of periodic waves of finite amplitude on the surface of a deep fluid. *Journal of Applied Mechanics and Physics*, 2:190–194, 1968. DOI: 10.1007/bf00913182 123

[14] V. E. Zakharov and A. B. Shabat. Exact theory of two-dimensional self-focusing and one-dimensional self-modulation of waves in nonlinear media. *Soviet Physics–JETP*, 34:62–69, 1972. 128, 131

CHAPTER 7

Resonant Interaction Between Waves

In the previous chapter, we investigated the phenomenon that results from non-linear interaction with itself when there is a single wavetrain. In this chapter, we will learn about the resonant interaction that occurs when multiple wavetrains coexist. If the wavenumber and frequency of three wavetrains satisfy a condition called the "resonance condition," interesting phenomena appear such as automatic generation of the third wave from the state where there are only two waves.

7.1   THREE-WAVE INTERACTION

7.1.1   BOUND WAVE COMPONENT

In Chapter 6, we studied that, when there is a single wavetrain, the complex amplitude $a(x, t)$ changes in the time scale of T_0/ϵ^2 and the space scale of about L_0/ϵ^2 as a result of nonlinear interactions with itself, and that its evolution is governed by the nonlinear Schrödinger equation (6.33). Here, ϵ is a dimensionless parameter that indicates the smallness of amplitude, and T_0 and L_0 are the typical period and wavelength of the carrier wave, respectively.

On the other hand, when multiple wavetrains coexist, nonlinear effects may become noticeable in shorter time and spaces scales. Suppose that a certain physical quantity $u(x, t)$ consists of two component waves (we will call them fundamental waves) as

$$u(x,t) = \left(a_1 e^{i\theta_1} + \text{c.c.}\right) + \left(a_2 e^{i\theta_2} + \text{c.c.}\right), \quad \theta_i = k_i x - \omega_i t \quad (i = 1, 2) \tag{7.1}$$

in its lowest order $O(\epsilon)$. Then as can be seen from

$$u^2 = \left\{ a_1^2 e^{2i\theta_1} + a_2^2 e^{2i\theta_2} + a_1 a_2 e^{i(\theta_1+\theta_2)} + a_1 a_2^* e^{i(\theta_1-\theta_2)} \right\} + \text{c.c.} + 2\left(|a_1|^2 + |a_2|^2\right), \tag{7.2}$$

if the governing equation or the boundary condition of the system includes second-order nonlinear terms of $u(x, t)$, then wave components with new pair of wavenumbers and frequencies such as $(2k_1, 2\omega_1)$, $(2k_2, 2\omega_2)$, $(k_1 \pm k_2, \omega_1 \pm \omega_2)$ are automatically generated. However, in the case of waves that are truly dispersive (i.e., other than $\omega = c_0 k$), the combination of these wavenumbers and frequencies produced by nonlinearity generally does not satisfy the dispersion relation. These oscillatory components that do not satisfy the dispersion relation can only be present with the fundamental waves, and are called **bound wave**.[1] The bound wave is just an "appendix" or "slave" of the fundamental wave, and its amplitude remains of $O(\epsilon^2)$ as (7.2) implies and does not grow to the level of $O(\epsilon)$ equal to the fundamental wave components.

Here we will confirm this by numerical simulation for a system governed by a simple model equation

$$u_t + u u_x + \mathcal{L}[u] = 0, \tag{7.3}$$

where $\mathcal{L}[u]$ is a linear operator acting on $u(x, t)$ that gives a dispersion relation

$$\omega(k) = \sqrt{k + k^3}, \tag{7.4}$$

which is the only requirement for $\mathcal{L}[u]$ to meet. As shown in Chapter 3, the linear dispersion relation of water surface waves considering both gravity and surface tension as restoring forces is given by (3.35). When the water depth is deep enough (i.e., $\tanh kh = 1$) and if the wavenumber k and frequency ω are made dimensionless with appropriate representative values, (3.35) reduces to the form of (7.4). Also, the nonlinear term of (7.3) is a typical one that appears for long waves, as seen in the KdV equation, etc. in Chapter 5. Therefore, it can be said that (7.3) is a model equation that has both the dispersion of the capillary-gravity type and the nonlinearity of the long wave type.

Figure 7.1 shows the result of a numerical simulation of (7.3)[2] in the case where the initial condition consists of superposition of "wave 1" with $(k_1, a_1) = (0.3, 0.01)$ and "wave 2" with $(k_2, a_2) = (1.0, 0.01)$. In the figure, the time evolution of the square of the amplitude (which is directly proportional to the energy) of the fundamental waves k_1, k_2 and the bound wave components $k_p = k_1 + k_2 (= 1.3)$ and $k_m = k_2 - k_1 (= 0.7)$ are shown. We can see that the bound waves stay so small almost on the horizontal axis $E = 0$, and that most of the energy continues to be possessed by the two fundamental waves. When we eliminate the nonlinear term $u u_x$ and linearize the equation, the two fundamental waves will continue to have the initially given energy permanently without interfering with each other, and no bound waves will be generated at all. The result shown in Fig. 7.1 is close to such a linear situation, so in this case it can be said that nonlinearity does not plays any important roles.

Another result of numerical simulation of (7.3) is shown in Fig. 7.2. This time we choose $(k_1, a_1) = (0.3, 0.01)$, $(k_2, a_2) = (1.6, 0.01)$, that is, the only difference from the above case is that the wavenumber k_2 of "wave 2" is 1.6 instead of 1.0. However, as you can see, the behavior has become surprisingly different just by this change of combination of wavenumbers. It can be seen that, at the beginning, the energy of wave 2 decreases significantly, and instead the energy

---

[1]Although the name contains "waves," it is not a "wave" in the strict sense because it does not satisfy the dispersion relation that should be satisfied as a wave.

[2]There are many ways to solve (7.3) numerically, but here we use the simple method called the "split step method." In this method, in the time evolution of u, the part given by the linear term $\mathcal{L}[u]$ and the part given by the nonlinear term $u u_x$ are treated separately, and these two substeps are combined to obtain a complete one temporal step. The calculation of linear term is performed in the Fourier space by using fast Fourier transform. Refer to textbooks on numerical computation for general information on numerical methods including this specific method.

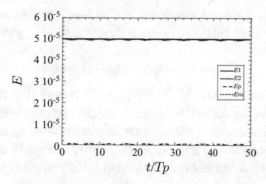

Figure 7.1: Evolution of energy of each component wave (non-resonance).

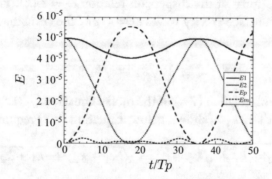

Figure 7.2: Evolution of energy of each component wave (resonance).

of the bound wave of wavenumber $k_1 + k_2 (= 1.9)$ increases, and at around 11 periods in terms of the period of wave 1, the bound wave becomes larger than wave 1 and becomes most energetic of all the wave components. However, after that, energy starts to return to wave 2 again, and the whole system almost returns to the initial state at around 35 periods. After that, active energy exchange between the three waves including this bound wave is repeated periodically. Why does such a big difference occur as seen in Figs. 7.1 and 7.2, although we only changed the wavenumber k_2 of wave 2?

7.1.2   THREE–WAVE RESONANT INTERACTION

As described above, in the case of dispersive waves, for most combinations of k_1 and k_2 of the fundamental waves, combinations of wavenumbers and frequencies generated by nonlinear interaction, i.e., $(2k_1, 2\omega_1)$, $(2k_2, 2\omega_2)$, $(k_1 \pm k_2, \omega_1 \pm \omega_2)$ do not satisfy the dispersion relation, so these components remain small as bound waves of the fundamental waves. However, depending on the combination of k_1 and k_2, the combination of wavenumber and frequency of these bound waves may satisfy the dispersion relation. In such a case, these components are no longer the bound wave of the fundamental waves, but are authentic waves, and actively participate in the process of energy exchange between waves on the equal footing with the fundamental waves. For example, suppose that

$$k_1 + k_2 = k_3, \quad \omega_1 + \omega_2 = \omega_3 \tag{7.5}$$

holds simultaneously, where ω_i stands for $\omega(k_i)$. (7.5) means that the "sum component" of wave 1 and wave 2 satisfies the dispersion relation. In this case, if waves of wavenumbers k_1 and k_2 are initially given as fundamental waves, wave of wavenumber k_3 and frequency ω_3 is generated by nonlinear interaction and grows spontaneously. In fact, in the numerical calculation of Fig. 7.2, a special combination of k_1, k_2 was intentionally selected so that such a situation is realized.

Whether or not there exist combinations of three wavenumbers and frequencies that satisfy (7.5) depends on the form of the dispersion relation $\omega = \omega(k)$. For example, for the dispersion relation of water surface gravity waves $\omega = \sqrt{gk}$, the combination of such three waves does not exist because

$$\omega_1 + \omega_2 = \omega_3 \quad \longrightarrow \quad \sqrt{k_1} + \sqrt{k_2} = \sqrt{k_1 + k_2} \quad \longrightarrow \quad \sqrt{k_1 k_2} = 0. \tag{7.6}$$

However, for the dispersion relation (7.4) of the model equation (7.3) considered here, there are pairs that satisfy (7.5) for any k_1 as shown below. Equation (7.5) requires

$$\sqrt{k_1 + k_1^3} + \sqrt{k_2 + k_2^3} = \sqrt{(k_1 + k_2) + (k_1 + k_2)^3}, \tag{7.7}$$

which gives a cubic equation for k_2 as follows:

$$9k_1 k_2^3 + \left(14k_1^2 - 4\right) k_2^2 + 9k_1^3 k_2 - 4\left(1 + k_1^2\right) = 0. \tag{7.8}$$

This cubic equation has one real root k_2 (> 0) for any k_1 (> 0), as shown in Fig. 7.3. The combination of the wavenumbers of the two fundamental waves adopted in the calculation of Fig. 7.2 is almost on this curve, and therefore the three waves with wavenumbers k_1, k_2, k_3 ($= k_1 + k_2$) are a special set that almost satisfies (7.5).

Generally, when the three wavenumbers k_1, k_2, k_3 and their corresponding frequencies $\omega_1, \omega_2, \omega_3$ satisfy

$$k_1 \pm k_2 \pm k_3 = 0, \quad \omega_1 \pm \omega_2 \pm \omega_3 = 0, \tag{7.9}$$

active energy exchange as described above occurs between these three waves. This phenomenon is called the **three-wave resonance** and (7.9) is called the **three-wave resonance condition**.

[Supplement]

When the target physical quantity takes only real values, a wave whose wavenumber, frequency and complex amplitude are (k, ω, a) is expressed as a sum with

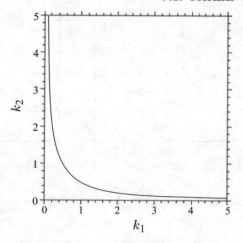

Figure 7.3: k_1 and k_2 satisfying the resonance condition (7.5).

a complex conjugate as $a\,e^{i(kx-\omega t)} + $ c.c.. This expression can also be written as $a^*\,e^{i[(-k)x-(-\omega)t]} + $ c.c.. Thus, a wave with (k,ω,a) and a wave with $(-k,-\omega,a^*)$ are identical. From this, it can be seen that k can be negative, and that ω should be extended as an odd function when extending from the region of $k > 0$ to that of $k < 0$.

However, this does not necessarily mean that $\omega(k)$ looks like an odd function. For example, for the equation $u_{tt} + u_{xxxx} = 0$, the dispersion relation is $\omega^2 = k^4$, that is, $\omega = \pm k^2$ and does not seem like an odd function. In this system, there are two branches (wave modes) $\omega = k^2$, $\omega = -k^2$ for one $k\,(> 0)$. When this is extended to $k < 0$, it should not be understood that $\omega = k^2$ and $\omega = -k^2$ are connected on both sides across $k = 0$ as even functions. It should be understood that $\omega = k^2$ for $k > 0$ is connected to $\omega = -k^2$ for $k < 0$ while $\omega = -k^2$ for $k > 0$ is connected to $\omega = k^2$ for $k < 0$ both as odd functions. In addition to the above, considering that the numbering of waves is arbitrary, (7.5) alone includes all the cases of general resonance condition (7.9). However, for the sake of simplicity here, we will continue assuming that k and ω are positive.

We can also use graphs to see if the dispersion relation $\omega = \omega(k)$ allows pairs of k satisfying (7.5).

Suppose that the dispersion relation is represented by a curve as shown in Fig. 7.4a, and we will call this the dispersion curve. Suppose that, when the dispersion curve is copied on a transparent paper T and its origin is translated to the point (k_1, ω_1) on the original dispersion curve, the two dispersion curves intersect at point A as shown in Fig. 7.4b. Let the coordinates of A on the transparent paper T be (k_2, ω_2) and the coordinates of the point on the original

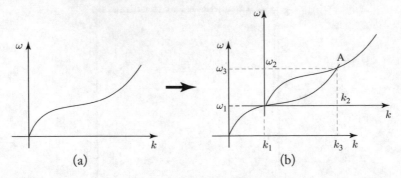

Figure 7.4: Graphical method of finding a set of k satisfying the resonance condition (7.5).

graph be (k_3, ω_3), then it is clear from the method of construction that a three-wave resonance condition (7.5) holds between the three (k_i, ω_i) $(i = 1, 2, 3)$. In this way, we can see whether sets of k satisfying the three-wave resonance condition exist, and we can also know their rough positions if they exist.

7.2   THREE–WAVE INTERACTION EQUATION

7.2.1   DERIVATION OF THREE–WAVE INTERACTION EQUATION

In order to gain a more quantitative understanding of the energy exchange between the three waves that satisfy the resonance condition, taking the model equation (7.3) as an example, let us derive by multiple scale method the equations governing the time evolution of the complex amplitude of the three waves. The basic idea is almost the same as when the NLS equation was derived in the previous chapter. First, the dependent variable $u(x, t)$ is expanded in a small parameter ϵ as

$$u(x,t) = \sum_{j=1} \epsilon^j u_j = \epsilon u_1 + \epsilon^2 u_2 + \cdots, \quad u_j = u_j(x_0, t_0, t_1, t_2, \cdots), \quad (j = 1, 2, \ldots).$$

(7.10)

Here, x_0, t_0 are the fast variables that look at each wave in the wavetrain, $t_1 = \epsilon t_0$ is a slowly changing time variable to handle a long time when nonlinear effects become apparent, and $t_2 = \epsilon^2 t_0$ represents a more slowly changing time variable to see higher-order nonlinear effects. Unlike in the case of the derivation of the NLS equation, however, we do not introduce slow spatial variables x_1, x_2 to handle the effects of higher-order dispersion, i.e., the effects resulting from having a narrow but finite-width spectrum of each wavetrain. As the time variable t is extended to multiple $t_i's$, the time derivative is expanded as

$$\frac{\partial}{\partial t} = \frac{\partial}{\partial t_0} + \epsilon \frac{\partial}{\partial t_1} + \epsilon^2 \frac{\partial}{\partial t_2} + \cdots.$$

(7.11)

Substituting the expansions (7.10) and (7.11) into (7.3) and sorting them out according to the power of ϵ, an approximate solution can be obtained as follows by solving them from the lower order.

First, the problem of $O(\epsilon)$ gives

$$\frac{\partial u_1}{\partial t_0} + \mathcal{L}[u_1] = 0. \tag{7.12}$$

Here, we adopt the sum of three waves that satisfy the three-wave resonance condition (7.5) as the solution of the lowest order $O(\epsilon)$. That is,

$$u_1 = a_1(t_1, t_2, \cdots)e^{i\theta_1} + a_2(t_1, t_2, \cdots)e^{i\theta_2} + a_3(t_1, t_2, \cdots)e^{i\theta_3} + \text{c.c.}, \tag{7.13a}$$

$$\theta_j = k_j x_0 - \omega_j t_0, \quad \omega_j = \sqrt{k_j + k_j^3} \quad (j = 1, 2, 3), \tag{7.13b}$$

$$k_1 + k_2 = k_3, \quad \omega_1 + \omega_2 = \omega_3. \tag{7.13c}$$

From (7.13c), $\theta_1 + \theta_2 = \theta_3$ also holds. As in the case of the modulation problem of a quasi-monochromatic wavetrain, the complex amplitude a_j of each wavetrain is constant with respect to the fast time t_0, but it is allowed to depend on slow time variables t_1 and t_2.

The next $O(\epsilon^2)$ problem is given as follows:

$$\frac{\partial u_2}{\partial t_0} + \mathcal{L}[u_2] = -\frac{\partial u_1}{\partial t_1} - u_1 \frac{\partial u_1}{\partial x_0} = -\left(\frac{\partial a_1}{\partial t_1} e^{i\theta_1} + \frac{\partial a_2}{\partial t_1} e^{i\theta_2} + \frac{\partial a_3}{\partial t_1} e^{i\theta_3} + \text{c.c.} \right)$$
$$- \left(a_1 e^{i\theta_1} + a_2 e^{i\theta_2} + a_3 e^{i\theta_3} + \text{c.c.} \right)\left(ik_1 a_1 e^{i\theta_1} + ik_2 a_2 e^{i\theta_2} + ik_3 a_3 e^{i\theta_3} + \text{c.c.} \right). \tag{7.14}$$

The nonlinear terms on the right-hand side are somewhat complicated, but from there come out terms like $e^{2i\theta_1}$, $e^{2i\theta_2}$, $e^{2i\theta_3}$, $e^{i(\theta_3 \pm \theta_1)}$, $e^{i(\theta_3 \pm \theta_2)}$, $e^{i(\theta_2 \pm \theta_1)}$ and their complex conjugates. Considering that $\theta_1 + \theta_2 = \theta_3$ holds, the following also hold:

$$e^{i(\theta_3 - \theta_1)} = e^{i\theta_2}, \quad e^{i(\theta_3 - \theta_2)} = e^{i\theta_1}, \quad e^{i(\theta_1 + \theta_2)} = e^{i\theta_3}. \tag{7.15}$$

Here, $e^{i\theta_1}$, $e^{i\theta_2}$, $e^{i\theta_3}$ are the homogeneous solutions of (7.14), that is, the solutions of the equation when the right side is 0. As already mentioned in the perturbation method in Chapter 4 and the derivation example of the NLS equation in Section 6.3.3, if such a homogeneous solution is present on the right side as a forcing term, resonance occurs and a secular term appears, and as a result, the perturbation method breaks down. Including the part that comes out of the term $\frac{\partial u_1}{\partial t_1}$, the part on the right side which is proportional to $e^{i\theta_1}$ is

$$\left[-\frac{\partial a_1}{\partial t_1} - i(k_3 - k_2)a_2^* a_3 \right] e^{i\theta_1} = \left(-\frac{\partial a_1}{\partial t_1} - ik_1 a_2^* a_3 \right) e^{i\theta_1}. \tag{7.16}$$

Then, from the non-secular condition that the coefficients of $e^{i\theta_1}$, $e^{i\theta_2}$, $e^{i\theta_3}$ on the right side become all 0, we obtain the following set of equations that govern the time evolution of the complex amplitude of the three wavetrains that satisfy the resonance condition (7.5):

$$\dot{a}_1 = -ik_1 a_2^* a_3, \quad \dot{a}_2 = -ik_2 a_1^* a_3, \quad \dot{a}_3 = -ik_3 a_1 a_2, \tag{7.17}$$

where $\dot{a}_j$ on the left side denotes the time derivative of a_j.

In the above, we have derived (7.17) for a specific system governed by the model equation (7.3). However, if we apply the same analysis for an arbitrary system with a dispersion relation that allows three-wave resonance, we almost always obtain a set of equations as follows, except in exceptional cases,

$$\dot{a}_1 = -i\gamma_1 a_2^* a_3, \quad \dot{a}_2 = -i\gamma_2 a_1^* a_3, \quad \dot{a}_3 = -i\gamma_3 a_1 a_2. \tag{7.18}$$

This is the standard equation that describes the interaction of three wavetrains that satisfy the three-wave resonance condition of the form (7.5), and we will call it the **three-wave interaction equation**.[3] The coefficient γ_j is a real number in a system that conserves energy as the system we are treating here in this chapter.

7.2.2   MANLEY–ROWE RELATIONS

If we introduce

$$N_1 = \text{sgn}(\gamma_1)|\gamma_2\gamma_3||a_1|^2, \quad N_2 = \text{sgn}(\gamma_2)|\gamma_3\gamma_1||a_2|^2, \quad N_3 = \text{sgn}(\gamma_3)|\gamma_1\gamma_2||a_3|^2, \tag{7.19}$$

in the three-wave interaction equation (7.18), we can see that

$$\frac{d}{dt}[N_1 - N_2] = 0, \quad \frac{d}{dt}[N_1 + N_3] = 0, \quad \frac{d}{dt}[N_2 + N_3] = 0, \tag{7.20}$$

which are known as **Manley–Rowe relations**. These relations indicate that the quantities in $[\cdots]$ do not change with time.

Also, if E_j and M_j are defined in terms of N_j by

$$E_j = \omega_j N_j, \quad M_j = k_j N_j, \quad (j = 1, 2, 3), \tag{7.21}$$

and considering the resonance condition (7.5) and the Manley–Rowe relations (7.20), it can be shown that

$$\frac{d}{dt}[E_1 + E_2 + E_3] = 0, \quad \frac{d}{dt}[M_1 + M_2 + M_3] = 0. \tag{7.22}$$

N_j, E_j, M_j often have physical meanings of wave action, energy and momentum of each wavetrain, respectively, and in that case (7.22) corresponds to the conservation laws of energy and momentum.

EXAMPLE 1: CHECKING MANLEY–ROWE RELATIONS

Verify the Manley–Rowe relations (7.20).

---

[3]Here, we have not taken into account of the spectral width of each wavetrain. If we take this into consideration, it only replaces the time derivative of the left side of (7.18) with the translation at the group velocity of each wavetrain like $\frac{da_j}{dt} \rightarrow \frac{\partial a_j}{\partial t} + \omega'(k_j)\frac{\partial a_j}{\partial x}$.

[Answer]

From the definition (7.19) of N_1, N_2, and the three-wave interaction equation (7.18),

$$\frac{dN_1}{dt} = \text{sgn}(\gamma_1)|\gamma_2\gamma_3| \left(\dot{a}_1 a_1^* + \text{c.c.}\right) = \text{sgn}(\gamma_1)|\gamma_2\gamma_3| \left(-i\gamma_1\, a_2^* a_3\right) a_1^* + \text{c.c.}$$
$$= -i\,|\gamma_1\gamma_2\gamma_3|a_1^* a_2^* a_3 + \text{c.c.}, \tag{7.23a}$$
$$\frac{dN_2}{dt} = \text{sgn}(\gamma_2)|\gamma_3\gamma_1| \left(\dot{a}_2 a_2^* + \text{c.c.}\right) = \text{sgn}(\gamma_2)|\gamma_3\gamma_1| \left(-i\gamma_1\, a_1^* a_3\right) a_2^* + \text{c.c.}$$
$$= -i\,|\gamma_1\gamma_2\gamma_3|a_1^* a_2^* a_3 + \text{c.c.}, \tag{7.23b}$$

where we used $\text{sgn}(\gamma)\gamma = |\gamma|$. This proves the first equation of (7.20). The other two equations can also be shown to hold similarly.   ♣

. . . .

Although details are not described here, (7.18) can be reduced to a single equation that describes the motion of a point mass in a cubic potential by using the Manley–Rowe relations. Therefore, it can be solved analytically using Jacobi's elliptic functions [3]. According to it, when the signs of coupling coefficient γ_j are all the same, the analytical solution expresses the periodic energy exchange among three resonant wavetrains as we saw in Fig. 7.2.

On the other hand, if the signs of γ_j are not all the same, the analytical solution of (7.18) shows that the solution diverges to infinity in a finite time. Divergence of amplitude apparently contradicts the energy and the momentum conservation laws (7.22). However, the definition (7.21) of energy E_j includes $\text{sgn}(\gamma_j)$ so that the energy E_j of the jth wavetrain is negative if $\text{sgn}(\gamma_j) < 0$. Under such circumstances, it is not impossible for each E_j to diverge to positive or negative infinity while keeping the sum of E_j constant. However, do waves with negative energy ever exist in reality? Such a situation does not usually occur for waves in stationary fluid, which is the main focus of this book. However, for a broader wave phenomena, a seemingly strange wave called **negative energy wave** actually exists.

The concept of "the sign of energy of a wave" was originally developed for the wave phenomena in plasma physics. When trying to generate a wavetrain in a wave-less quiescent state, if it is necessary to supply some additional energy to the system, the energy of the wave is positive, if conversely it is necessary to withdraw some amount of energy from the system, the energy of the wave is said to be negative. In other words, if wave excitation reduces the total energy of the system, the wave is a negative energy wave. Some kind of waves in plasma have negative energy. Therefore, in the three-wave resonant interaction that involves such waves, the amplitudes of all three waves can become simultaneously infinite in finite time. This phenomenon has long been known in plasma physics and is called the **explosive instability**. Also in hydrodynamics, in a two-layer fluid system consisting of two fluid layers with different densities, it is known that the waves traveling along the interface become negative energy wave in certain situations when the two layers flow at different speeds [2].

7.3   WAVE GENERATION AND EXCITATION BY THREE–WAVE RESONANCE

Continuing from the previous section, we will look specifically at two interesting phenomena brought about by the three-wave resonant interactions.

7.3.1   GENERATION OF THE THIRD WAVE BY TWO WAVES

As considered at the beginning of this chapter, let us assume that only two waves, "wave 1" and "wave 2" exist initially, that is, at $t = 0, a_1 = a_{10} \neq 0, a_2 = a_{20} \neq 0, a_3 = 0$. Then, in the initial stage where a_3 is still very small and can be approximated as $a_3 \approx 0$, we can get from (7.18)

$$\dot{a}_1 \approx 0 \longrightarrow a_1(t) \approx a_{10}, \qquad \dot{a}_2 \approx 0 \longrightarrow a_2(t) \approx a_{20},$$
$$\dot{a}_3 \approx -i\gamma_3 a_{10}a_{20} \longrightarrow |a_3| \approx G\,t, \quad G = |\gamma_3 a_{10}a_{20}|. \tag{7.24}$$

From this, it is predicted that "wave 3" with the wavenumber $k_3 (= k_1 + k_2)$ which did not exist initially is generated by the resonant interaction, and its amplitude grows in proportion to time. Let us verify this by direct numerical simulation of the model equation (7.3) same as that shown in Fig. 7.2. However, since we are interested in the initial behavior of wave 3 this time, we will reduce the amplitudes of wave 1 and wave 2 to $a_{10} = 0.001$, $a_{20} = 0.001$ to make the initial stage of time evolution longer.

In the case of the three-wave interaction equation (7.17) for the model equation (7.3), $\gamma_3 = k_3 = 0.3 + 1.6 = 1.9$. Therefore, according to (7.24), the growth rate G of a_3 is predicted to be 1.9×10^{-6}. Figure 7.5 shows $|a_3|$ obtained from the numerical simulation of the model equation as a function of t, and it is clear that a_3 grows in proportion to t. In addition, the growth rate G estimated from the slope of the graph is $G = 1.83 \times 10^{-6}$, showing a good agreement with the theoretical value $G = 1.90 \times 10^{-6}$. It should be noted here that this numerical calculation directly simulates the model equation (7.3) itself, not the three-wave interaction equation (7.17). Therefore, it can be said that the good agreement between the theoretical prediction and the numerical simulation results on the initial growth rate G of wave 3 also indicates the validity of the three-wave interaction equation (7.17) itself.

Figure 7.6 shows the result of exactly the same numerical simulation as Fig. 7.5 for a longer time. The growth of a_3 in proportion to t seen in Fig. 7.5 holds only under the assumption that a_3 is small, and the behavior disappears as it grows and approaches the same size as a_1 and a_2. Looking at longer times, the periodic energy exchange between resonant three waves, which is the behavior predicted by the analytical solution of the three-wave interaction equation, can be clearly observed also in the direct numerical simulation of the model equation. Here we treated the case where wave 1 and wave 2 existed initially and wave3 with $k_3 (= k_1 + k_2)$ was generated and grew, but this phenomenon occurs in the same way when any two of the three resonating waves exist initially.

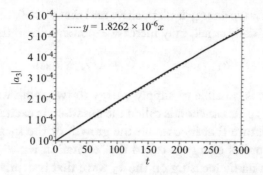

Figure 7.5: Generation and evolution of wave 3 by three-wave resonance (linear growth in initial stage).

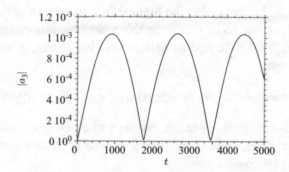

Figure 7.6: Generation and evolution of wave 3 by three-wave resonance (periodic behavior in long term).

7.3.2 EXCITATION OF TWO WAVES BY ONE WAVE

Then what happens if there is only one wave initially? According to the three-wave interaction equation (7.18), if there is only one wave, $\dot{a}_j = 0$ ($j = 1, 2, 3$) for all three waves, so no change occurs. However, as we will see below, if the other two waves have little but non-zero energy, they may grow rapidly. Suppose, for example, that most of the energy is concentrated in wave3 initially, that is, $|a_3| \gg |a_1|, |a_2| \approx 0$ at $t = 0$. Then $\dot{a}_3 \approx 0$ from the third equation of (7.18), hence $a_3 \approx a_{30}$ (constant). Differentiating the first and the second equations of (7.18) once again with respect to t with $a_3 \approx a_{30}$ in mind, we obtain

$$\ddot{a}_1 \approx -i\gamma_1 \dot{a}_2^* a_{30} = -i\gamma_1 (i\gamma_2 a_1 a_{30}^*) a_{30} = \gamma_1\gamma_2 |a_{30}|^2 a_1, \tag{7.25a}$$

$$\ddot{a}_2 \approx -i\gamma_2 \dot{a}_1^* a_{30} = -i\gamma_2 (i\gamma_1 a_2 a_{30}^*) a_{30} = \gamma_1\gamma_2 |a_{30}|^2 a_2. \tag{7.25b}$$

Since we assume that the wave energy is all positive and that $\gamma_j > 0$, this equation indicates that a_1 and a_2, which were initially small, may increase exponentially in time like

$$a_1, a_2 \propto e^{Gt}, \quad G = \sqrt{\gamma_1 \gamma_2} \, |a_{30}|. \tag{7.26}$$

This result implies that it is possible to supply energy to two weak waves (k_1, k_2) by injecting one strong wave (k_3). This phenomenon is called the **parametric excitation** or **decay instability**. The former is a nomenclature that focuses on the growth of the k_1 and k_2 waves, and in this case the k_3 wave that supplies energy is called the "excitation wave" or "pump wave," On the other hand, the latter is a name focusing on the k_3 wave that is deprived of energy. If the small amplitude waves of wavenumber k_1 and k_2 are regarded as small disturbances for a state in which only k_3 wave exists, the k_3 wave is naturally classified to be unstable in the sense that the small disturbances grow exponentially with time.

Considering the resonance condition (7.5), in the above case, the wave k_3 whose role is to supply energy to others is the wave with the highest frequency among the three waves. In fact, parametric excitation does not occur except in this situation. For example, suppose that only wave 1 has a large energy at $t = 0$, and let $|a_1| \gg |a_2|, |a_3|$. Then, if we do the same thing as deriving (7.25), we will get

$$a_1 \approx a_{10} \, (\text{constant}), \quad \ddot{a}_2 \approx -\gamma_2 \gamma_3 |a_{10}|^2 \, a_2, \quad \ddot{a}_3 \approx -\gamma_2 \gamma_3 |a_{10}|^2 \, a_3. \tag{7.27}$$

Since $\gamma_2 \gamma_3 |a_{10}|^2 > 0$, these equations mean that a_2 and a_3 behave oscillatory like cos and sin instead of exponentially. Thus, wave 1 or wave 2 cannot become a pump wave that supplies energy and excites other two waves.

All of the above discussions are not based on the governing equation of the system itself, in this case the model equation (7.3), but only theoretical predictions based on the approximate equations (7.18) derived by the perturbation method. In the following, we will verify the validity of the above predictions by performing direct numerical simulation of the model equation (7.3). We will employ $(k_1, k_2, k_3) = (0.3, 1.6, 1.9)$ again as a set of wavenumbers which almost satisfy the three-wave resonance condition (7.5). Figure 7.7 shows the time evolution of the energy of each wave when giving most energy to the wave of k_3 which has the highest frequency at $t = 0$ and the energy of about 1/100 is given to k_1 and k_2. As predicted by the three-wave interaction equation, it can be seen that the waves of k_1 and k_2 grow rapidly due to energy injection (pumping) from k_3. The periodic energy exchange seen later is also consistent with the behavior of the analytical solution of the three-wave interaction equation. According to the dispersion relation $\omega = \sqrt{k + k^3}$ of the model equation (7.3), the frequencies corresponding to $k_1 = 0.3, k_2 = 1.6, k_3 = 1.9$ are $f_1 = 0.09, f_2 = 0.38, f_3 = 0.47$, respectively. The result shown in Fig. 7.7 implies that a phenomenon something like the following can happen. Suppose that we generate waves by moving the wavemaker installed at one end of the wave flume at the frequency of 0.47 Hz, and observe the wave motion at a certain point downstream. Then, if we measure the wave at a point corresponding to "100" in the figure, there is almost no 0.47 Hz component that we had generated, but somehow waves with frequencies of 0.09 Hz and 0.38 Hz are prominently observed.[4] If this is the case of light, even though you inject a blue light with high frequency to the medium, it may change to a mixture of green and red when you take it out somewhere corresponding to the "100" in Fig. 7.7.

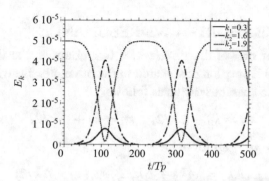

Figure 7.7: The evolution of energy of each wave when most of energy is given to k_3 initially.

On the other hand, Fig. 7.8 shows the result of the case when most of the energy is given to $k_1 = 0.3$ at $t = 0$. Since k_1 is not the wave with the highest frequency among the three resonant waves, it can be confirmed that pumping of k_2 and k_3 does not occur as predicted by the three-wave interaction equation, and the state in which only k_1 has energy continues over time. In the figure, the energy of k_2 and k_3 are too small to be distinguished from the horizontal axis.

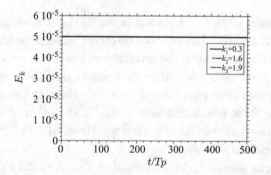

Figure 7.8: The evolution of energy of each wave when most of energy is given to k_1 initially.

---

[4]Note that this is a story of an imaginary world governed by the model equation, and exactly the same phenomenon does not happen with real water waves.

7.4   SPECIAL TYPES OF THREE–WAVE RESONANCE

Among the three-wave resonant interactions, there are two types with special names. One is the **long-wave short-wave resonance** and the other is the **harmonic resonance**. Let's look at each one below.

7.4.1   LONG-WAVE SHORT-WAVE RESONANCE

When the wavelength of one of the three waves (for example, the first wave) satisfying the resonance condition (7.5) is very long compared to the other, i.e., the wavenumber is very small, the condition for k of (7.5) can be written as follows:

$$k_1 = \Delta k, \quad k_2 = k_0 - \Delta k/2, \quad k_3 = k_0 + \Delta k/2, \quad (\Delta k \ll k_0). \tag{7.28}$$

Then the condition $\omega_1 = \omega_3 - \omega_2$ for ω gives

$$\omega(\Delta k) = \omega(k_0 + \Delta k/2) - \omega(k_0 - \Delta k/2), \tag{7.29}$$

this can be written as

$$\lim_{\Delta k \to 0} \frac{\omega(\Delta k)}{\Delta k} = \frac{d\omega}{dk}\bigg|_{k_0} \tag{7.30}$$

by dividing both sides by Δk and taking the limit of $\Delta k \to 0$. Note that the left side represents the phase velocity of the "long wave" with wavenumber Δk, while the right side represents the group velocity of the "short wave" with wavenumber k_0. That is, in the case of three-wave resonance including waves having very different wavelengths, the resonance condition can be expressed as "the group velocity of the short wave and the phase velocity of the long wave are equal."

In a strongly dispersive wave, the short waves and the long waves generally have very different propagation velocities, and they have almost no direct interaction with each other. However, a strong interaction may occur between the envelope of the short wave (such as modulation of the amplitude) which is transmitted at the group velocity and the waveform of the long wave by both propagating together at an equal speed for a long time. In this case, the original picture of the interaction among three resonant waves is modified to that of interaction between two components, i.e., interaction between a modulated wavetrain $a(x,t)e^{i(k_0 x - \omega_0 t)} + $ c.c. and a long and non-oscillatory fluctuation $B(x,t)$.

If we derive equations that describe the evolution of $a(x,t)$ and $B(x,t)$ from the governing equations of the system by perturbation method in the situation when long-wave short-wave resonance occurs, we often get a system of equations like follows:

$$i\left\{\frac{\partial a}{\partial t} + \omega'(k_0)\frac{\partial a}{\partial x}\right\} + p\frac{\partial^2 a}{\partial x^2} = q_1 a B, \qquad \frac{\partial B}{\partial t} + c_p \frac{\partial B}{\partial x} = q_2 \frac{\partial |a|^2}{\partial x}, \tag{7.31}$$

where c_p is the phase velocity of the long wave. The meaning of the left side of the first equation for the complex amplitude a of the short wave is the same as in the case of the NLS equation. The right side expresses an effect like a Doppler effect by which the frequency of k_0-wave changes by an amount proportional to B that appears to be "flow" rather than "wave" when viewed from the short wave k_0. On the other hand, the second equation for B shows that spatial modulation of the short wave $(|a|^2)_x$ works as the source for producing long wave component.

One example of this phenomenon is the interaction between the waves on the surface of the sea (short wave) and the internal wave (long wave) associated with density stratification [8]. In many sea areas, it is often observed that the sea water temperature changes rapidly near some water depth, and above this narrow transition layer is a seawater which is warmer and lighter, and below it is a seawater which is colder and heavier. Such an interface of density and temperature is called pycnocline or thermocline. In such a two-layer fluid, there exit two kinds of waves, the surface wave mode in which the water surface is mainly displaced and the interfacial wave mode (or internal wave mode) in which the interface is mainly displaced. In a typical situation, the wavelength of the interfacial wave propagating along the thermocline is about 100 m, and its phase velocity is about 50 cm/s. The phase velocity of the surface wave that can be resonant with this through the long-wave short-wave resonance is about 1 m/s, so that its wavelength is less than 1 m, which is rather short wave. When a long wavelength interfacial wave is generated by the movement of the seawater due to tides colliding with the undulation of the seabed, a short wavelength surface wave is generated on the water surface synchronizing with the phase of the long interfacial waves. Photographs of the sea surface taken from satellites and aircrafts often show regular patterns of light and dark stripes in the surface waves of the sea. This is considered to be a visualization of the modulation pattern of the surface waves which is generated by the interfacial waves through the long-wave short-wave resonance between the interfacial waves and the surface waves.[5]

7.4.2   HARMONIC RESONANCE

Another special type of three-wave resonance occurs when two of the three waves are the same wave. In the case of $k_1 = k_2 = k_0$, $k_3 = 2k_0$ in (7.5), the condition for ω becomes

$$\omega(2k_0) = 2\omega(k_0) \quad \longrightarrow \quad \frac{\omega(2k_0)}{2k_0} = \frac{\omega(k_0)}{k_0}. \tag{7.32}$$

This requires that the second harmonic of the fundamental wave k_0 has the same phase velocity as the fundamental wave. This type of three-wave resonance is called the **second-harmonic resonance**. When the lowest-order $O(\epsilon)$ solution consists of a single fundamental wavetrain $a_0 \, e^{i(k_0 x - \omega_0 t)}$, the nonlinearity generates the second harmonic component $a_0^2 \, e^{i(2k_0 x - 2\omega_0 t)}$. As discussed at the beginning of this chapter, if the system is dispersive, the combination $(k, \omega) = (2k_0, 2\omega_0)$ does not satisfy the dispersion relation in most cases, and this component usually stays as small as $O(\epsilon^2)$ as a bound wave of the fundamental wave (k_0, ω_0). However, when (7.32) is

---

[5]Many photographs of sea surface waves which show the existence of waves inside the ocean can be seen on several websites, for example at http://www.internalwaveatlas.com/.

satisfied, this second harmonic itself becomes an authentic wave that satisfies the dispersion relation, and participates in the energy exchange on equal footing with the fundamental wave.

The existence of the second harmonic resonance also becomes apparent in the process of calculating the Stokes wave solution of the capillarity gravity wave on the water surface. As discussed in Section 3.4.2, when finding a nonlinear correction to the sinusoidal wave solution

$$\eta(x,t) \sim a_1 \cos(kx - \omega t), \quad \omega(k) = \sqrt{gk + \frac{\tau}{\rho}k^3} \tag{7.33}$$

of the linear theory, we first assume this solution at the lowest-order $O(\epsilon)$ of the amplitude, and find nonlinear solution in an expansion form in ϵ. For most wavenumbers k, this perturbation expansion can proceed without problems, and the higher-order terms are found sequentially in the form of harmonics $a_n \cos[n(kx - \omega t)]$. In this case, these nth harmonic components are bound waves, and their amplitude a_n becomes $O(a_1^n)$ (Stokes expansion).

However, when $k = \sqrt{\rho g/2\tau}$, the denominator of the expression for the amplitude a_2 of the second harmonic becomes 0, and the perturbation calculation breaks down. According to the linear dispersion relation $\omega(k) = \sqrt{gk + \tau k^3/\rho}$ of capillary gravity waves, it can be seen that the condition (7.32) of second-harmonic resonance is satisfied when $k = \sqrt{\rho g/2\tau}$. In this case, the second harmonic that appears in the Stokes expansion is a "free wave" that itself satisfies the dispersion relation, and has the right to have energy equivalent to the fundamental wave through the second harmonic resonance.

Therefore, in this situation, in order to find a nonlinear approximate solution corresponding to the Stokes wave, it is necessary to start assuming the form

$$\eta(x,t) \sim a_1 \cos(kx - \omega t) + a_2 \cos(2kx - 2\omega t + \theta_2) \tag{7.34}$$

instead of the usual (7.33) as the lowest order solution. The steadily traveling wave solution obtained by such an analysis assuming the lowest order as (7.34) is known as **Wilton's ripple** in the context of water waves [6]. Wilton's ripple becomes a steadily traveling waves with somewhat complex waveforms with two peaks in one wavelength, as shown in Fig. 7.9.[6]

In Example 1 of Chapter 6, for a system governed by the KdV equation (6.34), it was confirmed that the NLS equation describing the evolution of the complex amplitude $a(x,t)$ of a quasi-monochromatic wavetrain is given in the form

$$i\left(\frac{\partial a}{\partial t} + v_g \frac{\partial a}{\partial x}\right) + \frac{1}{2}\frac{dv_g}{dk}\frac{\partial^2 a}{\partial x^2} = \alpha^2 \left\{\frac{k^2}{D(2k, 2\omega)} + \frac{k}{v_g - c}\right\} a|a|^2. \tag{7.35}$$

A point to be particularly noted here is that if the carrier wavenumber k satisfies $D(2k, 2\omega) = 0$ or $v_g(k) = c$, then the coefficient of the nonlinear term on the right side of (7.35) diverges.

---

[6]In the dispersion relation $\omega(k) = \sqrt{gk + (\tau/\rho)k^3}$, $\omega(nk_n)/nk_n = \omega(k_n)/k_n$ also holds for wavenumber k_n given by $k_n = \sqrt{\rho g/n\tau}$ $(n = 1, 2, 3, \ldots)$. As a result, the phase velocity of the nth harmonic becomes equal to that of the fundamental wave, and strong interaction occurs between them as well.

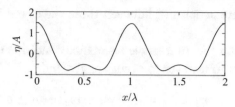

Figure 7.9: An example of the wave profile of Wilton's ripple.

Obviously, the former case corresponds to the second harmonic resonance, and the latter case corresponds to the long-wave short-wave resonance. As apparent from the process of derivation of (7.35), if the governing equation of the system includes a quadratic nonlinearity, then $D(2k, 2\omega)$ always appears in the denominator of the coefficient of the second harmonic component ($\propto a^2 e^{2i\theta}$) of $O(\epsilon^2)$. It is also a general property that the propagation of the DC component (= long wave) of $O(\epsilon^2)$ generated by nonlinear interaction is related to the amplitude modulation ($\propto \partial |a|^2/\partial x$) through the equation like (6.48a). Therefore, the fact that the factors $1/D(2k, 2\omega)$ and $1/(v_g - c)$ appear in the coefficient of the NLS equation is not limited to the case of using the KdV equation as the starting point but is a very common phenomenon. Thus, the three-wave resonant interaction may appear in the form of the harmonic resonance and the long-wave short-wave resonance even in a seemingly unrelated problems of modulation of quasi-monochromatic wavetrain, and sometimes requires a modification of the method of analysis.

7.5   FOUR–WAVE RESONANT INTERACTION

Depending on the form of the dispersion relation, there may be no set of three waves satisfying the three-wave resonance condition. The dispersion relation of surface gravity wave $\omega = \sqrt{gk}$ is one such example. In such a case, the resonant interaction between four waves is the most important nonlinear interaction. Phillips [7] was the first to study nonlinear interactions between waves in the context of fluid dynamics. He was a research student at that time at the University of Cambridge, England, where theoretical research on turbulence of fluid dynamics was actively carried out. One of the main concerns in turbulence theory was the energy transfer between different wavenumbers due to nonlinearity of the Navier–Stokes equation. After that, Phillips, who became interested in water waves, thought that "the governing equations of water waves are also nonlinear, and there should be energy transfer between different waves as in turbulence," and he started to calculate the nonlinear interaction between gravity waves. Ironically, the dispersion relation of gravity waves is a type that does not allow three-wave resonance, and he had to carry out very cumbersome analysis up to the third order of the amplitude in order to find a strong interaction between waves. Thus, the study of nonlinear interaction of water waves began with the study of the resonant interaction not between three waves but between higher-order four waves.

As can be inferred from the three-wave resonance condition (7.9), the condition for the resonant interaction between four waves to occur are give by[7]

$$\boldsymbol{k}_1 \pm \boldsymbol{k}_2 \pm \boldsymbol{k}_3 \pm \boldsymbol{k}_4 = 0, \quad \omega_1 \pm \omega_2 \pm \omega_3 \pm \omega_4 = 0. \tag{7.36}$$

In particular, in the case of the dispersion relation of gravity waves, it is known that, among various combinations of the signs in (7.36), there exist sets of four waves that satisfy the resonance condition only for

$$\boldsymbol{k}_1 + \boldsymbol{k}_2 = \boldsymbol{k}_3 + \boldsymbol{k}_4, \quad \omega_1 + \omega_2 = \omega_3 + \omega_4, \quad \omega_j = \sqrt{g|\boldsymbol{k}_j|} \quad (j = 1, \ldots, 4). \tag{7.37}$$

This always holds if $\boldsymbol{k}_j$ are all equal, or if $\boldsymbol{k}_1 = \boldsymbol{k}_3$, $\boldsymbol{k}_2 = \boldsymbol{k}_4$. Thus, unlike the three-wave resonance, sets of four waves that satisfy the four-wave resonance condition always exist regardless of the form of the dispersion relation.

Suppose that the lowest order water surface displacement can be written as

$$\eta(x, t) = \sum_{j=1}^{4} a_j(t) e^{i\theta_j} + \text{c.c.}, \quad \theta_j = \boldsymbol{k}_j \cdot \boldsymbol{x} - \omega_j t \tag{7.38}$$

as a sum of four gravity waves that satisfy the resonance condition (7.36). Then, as can be inferred from the derivation processes of the NLS equation (6.33) and that of the three-wave interaction equation (7.18), the evolution equations for the complex amplitude $a_j(x, t)$ are governed by a system of equations consisting of

$$\dot{a}_1 = i \left(\sum_{j=1}^{4} q_{1j} |a_j|^2 \right) a_1 + i\gamma_1 a_2^* a_3 a_4, \tag{7.39}$$

and three similar equations for $\dot{a}_j$ ($j = 2, 3, 4$) [1].[8] The coefficients q_{ij} and γ_i are all real numbers, and their values depend on the four wavenumber vectors $\boldsymbol{k}_j$. On the right side, the term $i \left(\sum_{j=1}^{4} q_{ij} |a_j|^2 \right) a_i$ indicate that the frequency of the ith wave is modified by an amount proportional to the square of the amplitude of the jth wave. The diagonal element q_{ii} is nothing but the correction resulting from the interaction with itself, i.e., the coefficient q of the nonlinear term of the NLS equation. On the other hand, the off-diagonal element q_{ij} ($i \neq j$) represents an effect similar to Doppler shift caused by the DC component of $O(\epsilon^2)$ that are generated as a result of self-interactions of other waves. It is known that these terms do not change the length

---

[7]Since water waves propagate in the 2D plane, i.e., the water surface, in ordinary circumstances, their wavenumbers are generally 2D vectors. Reflecting that, we here denote the wavenumber as a vector.

[8]When we also consider spatial modulation, the d/dt in (7.39) is replaced by $\partial/\partial t + \boldsymbol{v}_g(\boldsymbol{k}_j) \cdot \nabla$.

of a_i, and thus they only affect the wave speed of wave 1 but do not contribute to the exchange of energy. It is known that the energy exchange between four waves is brought about exclusively by the last term on the right hand side that includes γ_i. As can be seen from the form of (7.39), the rate of change of the complex amplitude a_i generated by the four-wave resonant interaction is of $O(\epsilon^3)$, so the space-time scale at which the effect appears prominently is T_0/ϵ^2, L_0/ϵ^2, which is much longer than three-wave resonance.

Most of the research on ocean waves before 1960 was based on the idea that ocean wave fields consist of linear waves and their superpositions, except for very specific topics such as Stokes waves. Even when Phillips proposed for the first time in 1960 the theory of energy exchange between gravity waves by four-wave resonant interactions, it seems that most researchers were quite skeptical about the importance of nonlinearity. In 1961, the U.S. Naval Oceanographic Office held a large international conference entitled "Ocean Wave Spectra." In its conference proceedings [9], hot controversies such as the one between Phillips and other researchers who had strong doubts about the possibility of creation of new waves by nonlinear interactions and of their growth to as large as $O(a)$ are recorded, which are all very interesting to read. Nowadays, the importance of nonlinear interactions is common sense, and is written in any textbook, but these controversies in the Proceedings suggest that only about 60 years ago, even the top researchers felt quite uncomfortable for the idea that such high-order nonlinear interactions generate waves and even change the energy spectra. However, such suspicions were completely wiped out by the two experiments conducted by the Longuet–Higgins group and the Phillips group [4, 5]. For ease of wave tank experiment, they chose a special set of wavenumbers satisfying (7.37) so that $k_1 = k_2$ and k_1 and k_3 are perpendicular to each other. They generated waves with wavenumber $k_1(= k_2)$ and k_3 from each of the wave makers installed on two adjacent sides of a rectangular wave tank, then they observed that an obliquely propagating wave with a wavenumber k_4 determined by (7.37) was actually generated, and also observed that the generated wave grew with the propagation distance at the growth rate expected from (7.39). These studies have established that the wave-wave interaction due to nonlinearity is an important factor for the development of the actual ocean waves and the evolution of the spectrum, and these ideas are also used in the numerical forecast of the ocean waves that are being conducted today.

7.6   REFERENCES

[1] D. J. Benney. Non-linear gravity wave interactions. *Journal of Fluid Mechanics*, 14:577–584, 1962. DOI: 10.1017/s0022112062001469 158

[2] R. A. Cairns. The role of negative energy waves in some instabilities of parallel flows. *Journal of Fluid Mechanics*, 92:1–14, 1972. DOI: 10.1017/s0022112079000495 149

[3] A. D. D. Craik. *Wave Interactions and Fluid Flows*. Cambridge U.P., 1985. DOI: 10.1017/cbo9780511569548 149

[4] M. S. Longuet-Higgins and N. D. Smith. An experiment on third-order resonant wave interactions. *Journal of Fluid Mechanics*, 25:417–435, 1966. DOI: 10.1017/s0022112066000168 159

[5] L. F. McGoldric, O. M. Phillips, N. E. Huang, and T. H. Hodgson. Measurements of third-order resonant wave interactions. *Journal of Fluid Mechanics*, 25:437–456, 1966. DOI: 10.1017/s002211206600017x 159

[6] L. F. McGoldrick. On wilton's ripples: A special case of resonant interactions. *Journal of Fluid Mechanics*, 42:193–200, 1970. DOI: 10.1017/s0022112070001179 156

[7] O. M. Phillips. On the dynamics of unsteady gravity waves of finite amplitude. part 1. the elementary interactions. *Journal of Fluid Mechanics*, 9:193–217, 1960. DOI: 10.1017/s0022112060001043 157

[8] O. M. Phillips. Wave interactions—the evolution of an idea. *Journal of Fluid Mechanics*, 106:215–227, 1981. DOI: 10.1017/s0022112081001572 155

[9] The U.S. Naval Oceanographic Office. *Ocean Wave Spectra, Proceedings of a Conference.* Prentice Hall Inc., 1963. 159

CHAPTER 8

Wave Turbulence: Interaction of Innumerable Waves

In the previous chapter, we considered about nonlinear interactions between three and four waves. However, for example, in the field of ocean waves generated by a storm or in the elastic waves generated when a thin iron plate is continuously struck, innumerable waves having different frequencies and propagation directions coexist. Such a state in which innumerable waves coexist and interact nonlinearly with each other is called "wave turbulence" or "weak turbulence." In this chapter, using ocean waves as a concrete example, we introduce a method of description of such a complex wave field in terms of statistical quantities such as representative wave heights and energy spectrum.

8.1 ENERGY SPECTRUM

When you stand on a shallow beach, you can see waves coming from offshore. As a wave approach the shore, its shape deforms, but you can keep watching one particular wave. The long waves targeted by the KdV equation of Chapter 5 are just such waves, and it would be natural to use the surface deformation $\eta(x, t)$ itself to describe the behavior of these waves. On the other hand, the waves you see when you get on a ship and go offshore are quite different. The shape of the water surface keeps changing irregularly without staying in the same shape even for a moment, and if you try to keep watching one particular wave, it disappears somewhere immediately. It is very difficult to handle such a complicated surface wave condition by treating directly the ever-changing surface deformation $\eta(x, t)$ itself, and it would not be practical even if it were possible.

Why do the ocean waves behave so complicated and irregularly? That is because an infinite number of wavetrains with different wavelengths and propagation directions overlap each other. In the case where many wavetrains with different wavenumbers overlap, surface displacement $\eta(x, t)$ can be written as

$$\eta(x,t) = \sum_{j=0}^{\infty} a_j \cos\left[k_j x - \omega(k_j)t + \theta_j\right] \tag{8.1}$$

if effects of small nonlinearity is ignored, where j is a subscript to distinguish each wavetrain. As shown in Chapter 3, according to the linear theory, the energy density E of a sinusoidal wave of amplitude a, i.e., the average energy that a sinusoidal wave has per unit length (per unit area if the propagation is in 2D), is given by $E = \rho g \overline{\eta^2} = \rho g a^2 / 2$. Therefore, the energy density of the wave field represented by (8.1) is given by

$$E = \rho g \overline{\eta^2} = \frac{1}{2} \rho g \sum_{j=0}^{\infty} a_j^2, \tag{8.2}$$

where $\overline{\eta^2}$ denotes the spatial average

$$\overline{\eta^2} = \lim_{L \to \infty} \frac{1}{L} \int_0^L \{\eta(x)\}^2 \, dx. \tag{8.3}$$

When discussing ocean waves, the density of water ρ and the gravitation acceleration g can be considered as constants, so although dimensionally incorrect, $\overline{\eta^2}$ is often called "energy density." Below we follow this convention.

When every wavenumber is included, (8.2) becomes a sum with respect to k which is very close, and it is convenient to introduce a continuous function $E(k)$ of k and express (8.2) as an integral as

$$\overline{\eta^2} = \int_0^{\infty} E(k) \, dk. \tag{8.4}$$

Then such $E(k)$ is called the **wavenumber spectrum**. From (8.2) and (8.4),

$$E(k) \, dk = \frac{1}{2} \sum_k^{k+dk} a_j^2, \tag{8.5}$$

where the sum on the right-hand side is taken for all waves with k_j that satisfy $k < k_j < k + dk$. From its definition, $E(k)$ tells us how the total energy E per unit length (or area) is distributed in the wavenumber k space. The wavenumber k and the frequency ω are linked through the dispersion relation, so (8.4) can also be expresses as

$$\overline{\eta^2} = \int_0^{\infty} E(k) \frac{dk}{d\omega} \, d\omega = \int_0^{\infty} F(\omega) \, d\omega, \qquad F(\omega) \equiv E(k)/\omega'(k), \tag{8.6}$$

and the energy density can also be expressed as an integral with respect to ω. $F(\omega)$ tells us how the energy is distributed in the ω space, and is called the **frequency spectrum**. The relation corresponding to (8.5) becomes

$$F(\omega) \, d\omega = \frac{1}{2} \sum_{\omega}^{\omega+d\omega} a_j^2, \tag{8.7}$$

where the sum is taken over all waves with ω_j that satisfy $\omega < \omega_j < \omega + d\omega$. For a wave field where an infinite number of wavetrains coexist and the waveform behaves in a complex and irregular manner, understanding the spectra such as $E(k)$ and $F(\omega)$ and predicting their spatio-temporal developments become the central subjects of research.

8.2   STATISTICS ABOUT WAVE HEIGHT

8.2.1   DEFINITION OF INDIVIDUAL WAVES AND REPRESENTATIVE WAVE HEIGHT

The Pierson–Moskowitz spectum (P-M spectrum):

$$F(\omega) = 5E\omega_m^4\omega^{-5}\exp\left[-\frac{5}{4}\left(\frac{\omega}{\omega_m}\right)^{-4}\right], \tag{8.8}$$

graphically shown in Fig. 8.1, is a representative frequency spectrum of a fully developed ocean wave field that has been blown by wind for a sufficiently long time and a long distance. Here, $E = \overline{\eta^2}$, and ω_m is the frequency at which $F(\omega)$ takes its maximum. Figure 8.2 shows an example of a time series which has the P-M spectrum as the frequency spectrum. Although this is an artificial time series created on a PC, it would be not so different from this if we plot the actual ocean wave observation data.

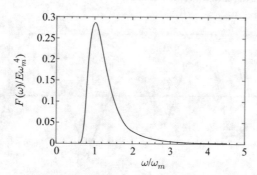

Figure 8.1: The Pierson–Moskowitz spectrum.

In the TV news program, the weather forecaster announces something like "Tomorrow's wave height in Tokyo Bay will be 2.0 m," but if you think of a complex waveform like Fig. 8.2, a question like "What does this wave height mean?" may come up. The wave height used in weather forecasting is a representative value of wave height called the "significant wave height" which is defined as follows.

In order to discuss the wave height, it is necessary to first divide the irregular waveform as shown in Fig. 8.2 into each single waves. For ocean waves themselves, however, there is no such thing that "This one wave starts at this point of time and ends at this point of time." Therefore,

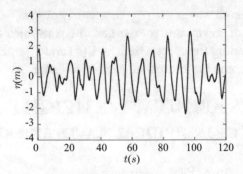

Figure 8.2: An example of time series corresponding to the P-M spectrum ($f_m = 10$ s).

no matter how we divide the continuous record of $\eta(t)$ into a collection of single wave, it can be nothing but quite artificial. One commonly used method for doing this is the **downward zero-crossing method**. Suppose there is a time record of water surface displacement $\eta(t)$ observed at a certain fixed point. In the downward zero-crossing method, as shown in Fig. 8.3, the time series is divided at the point where $\eta(t)$ crosses 0 downward and goes from positive to negative, and it treats that part between successive two downward zero-crossing points as one wave. Each wave thus defined is called an **individual wave**.[1]

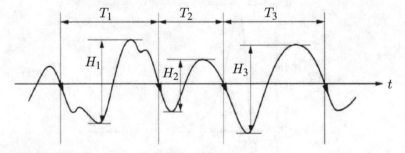

Figure 8.3: The downward zero-crossing method.

The time interval between the ends of each individual wave defines the "period" of that wave, and the difference between the maximum and minimum values of η between that time

[1]Of course, there is also a way called the **upward zero-crossing method**, which is opposite to this and divide the time series at points where $\eta(t)$ crosses 0 upward from negative to positive. When the target of the analysis is a time series, in the upward zero-crossing method, the wave crest is combined with the subsequent trough and interpreted as one wave, while in the downward zero-crossing method, the crest is combined with the preceding trough. There is usually no big difference in wave statistics, whichever we choose. However, in the case of visual observation, the wave height is evaluated by how high the wave crest is from the trough in front of it. Also, in the case where waves develop and lean forward and finally break, the front face of a high crest, that is, the slope between the crest and the trough before it becomes the most important part. From these facts, it seems that there are many researchers who prefer the downward zero-crossing method that treats the crest and the trough in front of it as a set [3].

defines the "wave height" of that individual wave. As you can imagine from Fig. 8.2, the wave heights of the individual waves defined in this way have different values. Assuming that 100 individual waves are included in the target time series, 100 different wave heights will appear. Then, 100 wave heights are arranged in the order from the largest one, and only one third of them from the largest (up to 33 in the present case) is taken out, and the average value of these is called the **significant wave height**, and is denoted by $H_{1/3}$. Also, the mean value of the period of these largest 1/3 individual waves is called the **significant wave period** and is denoted by $T_{1/3}$. The wave height forecasted in the TV program is this significant wave height calculated by such an average procedure.

8.2.2 PROBABILITY DISTRIBUTION OF WAVE HEIGHT

The significant wave height is a convenient statistical quantity that can roughly express the condition of the sea with only one quantity, but only with that information it is impossible to know how often a wave of a certain wave height appears. The probability density function (PDF) of wave height H is a function of H such that the probability that the wave height takes a value between H and $H + dH$ is express as $p(H)dH$. It is known that $p(H)$ is approximately given by

$$p(H) = \frac{\pi}{2} \frac{H}{\overline{H}^2} \exp\left[-\frac{\pi}{4}\left(\frac{H}{\overline{H}}\right)^2\right], \tag{8.9}$$

for a rather wide range of ocean conditions [9, 14]. Here $\overline{H}$ represents the average wave height defined by

$$\overline{H} = \int_0^\infty H\, p(H)\, dH. \tag{8.10}$$

The PDF (8.9) is called the **Rayleigh distribution**, which is one of the representative PDF that we can see in various fields of research, not limited to the wave height distribution of ocean waves.[2]

[Supplement]

In the process of theoretically deriving that the PDF of wave height H becomes (8.9), the spectrum is assumed to be narrow. However, as can be seen from the P-M spectrum shown in Fig. 8.1, the actual spectrum of ocean waves does not seem so narrow that such an approximation can be used. Nevertheless, this Rayleigh distribution holds fairly well for the actual ocean wave height distribution. One reason for this is that each individual wave is defined by the zero-crossing method. When

---

[2]It is known that the PDF of $\eta(t)$ becomes a Gaussian distribution according to the **central limit theorem** when $\eta(t)$ is a superposition of an infinite number of independent harmonic oscillations. And at this time, it can be shown theoretically that, in the limit where the width of the frequency spectrum of $\eta(t)$ is narrow, the PDF of the local maximum value of $\eta(t)$ (the value of η at the wave crest in the case of water surface displacement) is given by the Rayleigh distribution.

the spectrum is broad, the waves with higher frequencies (and hence shorter wave-lengths) appear in the waveform as being superimposed on the more energetic waves with lower frequency (and hence longer wavelength), hence these small waves tend to have little effect on the dividing process into individual waves in the zero-crossing method. For this reason, the zero-crossing method plays a role like a low-pass filter that allows only low-frequency components to pass, and is considered to have the effect of making the substantial spectrum width narrower than it actually is.

Note that if we introduce the dimensionless wave height $\xi = H/\overline{H}$ normalized by the average wave height $\overline{H}$, (8.9) can be written as

$$p(\xi) = \frac{\pi}{2}\,\xi\exp\left(-\frac{\pi}{4}\xi^2\right),\tag{8.11}$$

which is graphically shown in Fig. 8.4.

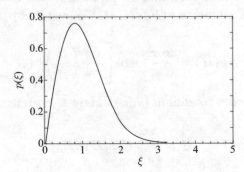

Figure 8.4: Rayleigh distribution for dimensionless wave height ξ.

Then, the "excess probability" $P(\xi)$ defined by $P(\xi) \equiv \int_\xi^\infty p(\xi')\,d\xi'$ (therefore $p(\xi) = -dP(\xi)/d\xi$) is given by $P(\xi) = \exp(-\pi\xi^2/4)$.

EXAMPLE 1: PDF OF DIMENSIONLESS WAVE HEIGHT

From the PDF of the wave height H given in (8.9), derive the PDF (8.11) of the dimensionless wave height ξ.

[Answer]

$$\int_0^\infty p(H)\,dH = \int_0^\infty p(\xi)\,d\xi = 1 \quad \longrightarrow \quad p(H)\,dH = p(\xi)\,d\xi.\tag{8.12}$$

From this and $H = \overline{H}\xi$,

$$p(\xi) = p(H)\frac{dH}{d\xi} = \frac{\pi}{2}\frac{\overline{H}\xi}{\overline{H}^2}\exp\left[-\frac{\pi}{4}\left(\frac{\overline{H}\xi}{\overline{H}}\right)^2\right]\overline{H} = \frac{\pi}{2}\xi\exp\left(-\frac{\pi}{4}\xi^2\right). \qquad (8.13)$$

♣

. . . .

If the wave height distribution $p(H)$ is known, the relationship between the average wave height $\overline{H}$ and the significant wave height $H_{1/3}$ can also be known. Let the wave height distribution $p(H)$ be given by (8.11), and let ξ_3 be the dimensionless wave height such that the probability of occurrence of waves with normalized wave height higher than ξ_3 is 1/3. Then from the definition of ξ_3

$$\int_{\xi_3}^{\infty} p(\xi)\,d\xi = \left[\exp\left(-\frac{\pi}{4}\xi^2\right)\right]_{\infty}^{\xi_3} = \exp\left(-\frac{\pi}{4}\xi_3^2\right) = \frac{1}{3}. \qquad (8.14)$$

From this, $\xi_3 = 2\sqrt{\ln 3}/\sqrt{\pi} \approx 1.183$. This means that, for example, when there are 300 waves, the wave height of the 100th wave counted from the largest wave height is 1.183 times the average wave height. Since the dimensionless wave height $\xi_{1/3}$ corresponding to the significant wave height $H_{1/3}$ is the average value of ξ larger than ξ_3, it is given by

$$\xi_{1/3} = \int_{\xi_3}^{\infty}\xi\, p(\xi)\,d\xi \bigg/ \int_{\xi_3}^{\infty} p(\xi)\,d\xi = \xi_3 + 3\,\mathrm{erfc}\left[\frac{\sqrt{\pi}}{2}\xi_3\right]. \qquad (8.15)$$

Here, $\mathrm{erfc}[x]$ is a function called "complementary error function," and is defined as follows together with the usual "error function" $\mathrm{erf}[x]$:

$$\mathrm{erf}[x] = \frac{2}{\sqrt{\pi}}\int_0^x e^{-t^2}\,dt, \quad \mathrm{erfc}[x] = \frac{2}{\sqrt{\pi}}\int_x^{\infty} e^{-t^2}\,dt = 1 - \mathrm{erf}[x]. \qquad (8.16)$$

Since $(\sqrt{\pi}/2)\xi_3 = \sqrt{\ln 3} \approx 1.048$, and $\mathrm{erfc}[1.048] \approx 0.138$, (8.15) gives,

$$\xi_{1/3} = \frac{H_{1/3}}{\overline{H}} \approx 1.183 + 3\times 0.138 = 1.597. \qquad (8.17)$$

That is, when the probability density of wave height is given by the Rayleigh distribution (8.9), the significant wave height is about 1.6 times the average wave height.

EXAMPLE 5: OCCURRENCE PROBABILITY OF FREAK WAVE

The freak wave mentioned in the previous chapter is often defined as a wave with a wave height exceeding twice the significant wave height. According to this definition, if it is predicted based on the Rayleigh distribution (8.9), how often does a freak wave appear?

[Answer]

Let H_f be the lowest wave height to be judged as a freak wave, and let ξ_f be the dimensionless wave height corresponding to H_f. Then from (8.17), $H_f = 2H_{1/3} = 3.194\overline{H}$, i.e., $\xi_f = 3.194$. Let P_f be the occurrence probability of freak waves,

$$P_f = \int_{H_f}^{\infty} p(H)\,dH = \int_{\xi_f}^{\infty} p(\xi)\,d\xi = \left[-\exp\left(-\frac{\pi}{4}\xi^2 \right) \right]_{\xi_f}^{\infty}$$
$$= e^{-8.012} = 3.31 \times 10^{-4}. \tag{8.18}$$

Therefore, a freak wave is expected to occur at a rate of about 1 wave in 3000 waves.[3]   ♣

. . . .

8.3   EVOLUTION EQUATION OF ENERGY SPECTRUM

In the case of the actual ocean waves, the wave travels in a horizontal 2D plane, and the wavenumber becomes a 2D vector $\boldsymbol{k}$. Along with that, the relation between $\overline{\eta^2}$ and energy spectrum $E(k)$ given by (8.4) is also rewritten to the integral on the 2D $\boldsymbol{k}$-plane as

$$\overline{\eta^2} = \int E(\boldsymbol{k})\,d\boldsymbol{k}. \tag{8.19}$$

The SMB method (see the column at the end of this chapter), the first practical wave forecast method developed from the need for military operations during World War II, aimed at predicting small number of statistical quantities such as significant wave height. However, thanks to the great strides of research and also of computation facilities, the object of prediction is now shifted to the vector wavenumber spectrum $E(\boldsymbol{k})$ which has huge amount of information. Although details are omitted here, in the modern wave forecasting method, $E(\boldsymbol{k})$ is considered to evolve spatially and temporally according to the following equation called the **energy balance equation**:

$$\frac{\partial E(\boldsymbol{k}; \boldsymbol{x}, t)}{\partial t} + \boldsymbol{v}_g(\boldsymbol{k}) \cdot \nabla_h E(\boldsymbol{k}; \boldsymbol{x}, t) = S_{nl} + S_{in} + S_{ds}. \tag{8.20}$$

(See, for example, [7, 12] and [4].) Here, ∇_h is the gradient operator in the horizontal xy-plane, so the left side merely expresses the result of the linear theory that the energy possessed by the wave with wavenumber $\boldsymbol{k}$ propagates at the corresponding group velocity $\boldsymbol{v}_g(\boldsymbol{k})$. On the other hand, the right side is the various source terms that cause the change of the spectrum. Among them, S_{nl} represents the energy exchange between different wavenumber components due to nonlinear interactions, S_{in} represents the energy input from wind, and S_{ds} represents energy dissipation due to breaking of overdeveloped waves. Although they are simply written as S as

---

[3]However, in the theory that uses Rayleigh distribution for the wave height distribution, it is assumed that the spectrum is narrow, and nonlinear effects are not considered. It is known that the fact that the actual ocean wave spectrum is not so narrow works in the direction to lower the occurrence probability of freak waves, while the effect of nonlinearity works in the direction to increase it.

a symbol in (8.20), they are actually rather complicated mathematical expressions including the spectrum $E(k)$ and other parameters such as wind speed and direction.

Of these three source terms, the wave breaking phenomenon that S_{ds} is trying to express is, in particular, a hydrodynamically extremely complex phenomenon of turbulent motion of two-phase flow which includes the interaction between sea water and air, generation of intense turbulent motion due to wave breaking, entrainment of bubbles into water, etc. Reflecting this, the modeling of S_{ds} has not yet reached the stage of being built on solid theory, and the current situation seems to be that S_{ds} is "tuned" so that the prediction results based on (8.20) become consistent with the huge observation data that are available nowadays by remote sensing from satellites, etc.

In contrast, for S_{nl} representing the energy exchange between waves due to nonlinear interactions, which is the subject of this book, there is a world standard theory developed by Hasselmann (1962) [2]. According to it, S_{nl} is expressed as

$$S_{nl}(k_4) = \iiint W_{1234}\delta(k_1 + k_2 - k_3 - k_4)\delta(\omega_1 + \omega_2 - \omega_3 - \omega_4)$$
$$\times \{E_1 E_2(E_3 + E_4) - E_3 E_4(E_1 + E_2)\}\, dk_1\, dk_2\, dk_3, \tag{8.21}$$

where abbreviations such as $E_1 = E(k_1)$, $\omega_1 = \omega(k_1)$ has been used. The coefficient W_{1234} is a complex function of k_1, k_2, k_3, k_4, and $\delta()$ is the Dirac delta function. Since the story is too complicated from the beginning, you do not need to worry at all if you do not understand this expression. However, there are two important points to be noted here. The first is that the integral expression giving S_{nl} contains two delta functions, one for k and another for ω. The delta function $\delta(x)$ takes on infinite value at $x = 0$, and remains 0 elsewhere, so it looks like an ultimately thin pulse. The fact that such delta functions are contained in the form of a product means that the exchange of energy occurs only between four waves for which

$$k_1 + k_2 - k_3 - k_4 = 0, \quad \omega_1 + \omega_2 - \omega_3 - \omega_4 = 0 \tag{8.22}$$

are satisfied at the same time, that is, the four wave resonance condition is satisfied. As mentioned earlier, for the dispersion relation of surface gravity waves $\omega = \sqrt{g|k|}$, there is no pair satisfying the three-wave resonance condition, and the four-wave resonance is the lowest order nonlinear resonant interaction that can be realized. We studied about the resonant interactions between three waves and four waves in the previous chapter. But even in the situation where there are innumerable many waves like the ocean wave field, the basic mechanism of energy exchange between them is still a resonant interaction between three waves, or four waves if three wave resonance is impossible.

Another important point to be noted is that not only in the case of surface water waves, but also in various kinds of wave phenomena, the rate of change of spectrum due to nonlinear interaction is generally given by an expression of the form similar to (8.21). As long as three-wave resonance do not exist and four-wave resonance is the main cause of spectral change, the evolution of the spectrum is generally described by an equation such as (8.21). The uniqueness or characteristics of each wave phenomenon such as whether the wave is a water wave, a wave traveling through an iron plate, or a wave in a plasma is exclusively appears in the concrete expressions of the dispersion relation $\omega(k)$ and the coupling coefficient W_{1234}. In a system where the dispersion relation allows three-wave resonance such as the model equation (7.3) in the previous section and capillary-gravity water wave, the rate of change of the spectrum S_{nl} due to nonlinear interaction is generally expressed something like

$$S_{nl}(k_3) = \iint \widetilde{W}_{123}\delta(k_1 + k_2 - k_3)\delta(\omega_1 + \omega_2 - \omega_3)\{E_1 E_2 - E_3(E_1 + E_2)\}\,dk_1 dk_2.$$
(8.23)

(See, for example, [10, 15].)

8.4   POWER LAW APPEARING IN ENERGY SPECTRUM

In the field observation of the frequency spectrum $F(\omega)$ of ocean wave, a spectrum of a power law such as $F(\omega) \propto \omega^{-4}$ is often reported in the frequency domain somewhat higher that the peak of the spectrum. For the readers who have read this book so far, this observational fact can be reasonably understood by using the dimension-based consideration and the method of roughly estimating the magnitudes of physical quantities that have often been used so far. So, I will take up the understanding this power-law spectrum as the final issue of this book. But before doing that, I will introduce as a preparation the **Kolmogorov spectrum** in turbulence, which is well known in fluid mechanics dealing with ordinary flows not wave motion.

8.4.1   KOLMOGOROV SPECTRUM OF TURBULENCE

What is turbulence?

Whether the flow of air around a moving car or the flow of water in a water pipe, the flow around us is very irregular and complicated, except when the flow speed is very slow. Figure 8.5 shows an example of temporal change of the wind speed measured at an observation tower set up in a wide field without influence of buildings. As you can see, the wind speed is constantly fluctuating irregularly. From the scale of the vertical axis, it can be seen that the data is not taken on a stormy day, but the wind speed is quite modest. This is a normal state for the wind speed. In fluid dynamics, such irregular flow is called a **turbulence**.

The motion of incompressible fluid is known to be governed by the Navier–Stokes equation (NS equation)

$$\frac{\partial v}{\partial t} + (v \cdot \nabla)v = -\frac{1}{\rho}\nabla p + \nu \nabla^2 v, \quad \nu = \frac{\mu}{\rho},$$
(8.24)

where μ and ν are the coefficient of viscosity and the kinematic viscosity, respectively, and both are material constants of the fluid. The NS equation is nonlinear, and energy exchange occurs between different wavenumber components. In the NS equation, it is the last viscosity term on

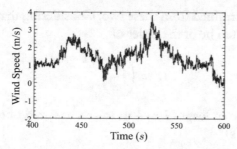

Figure 8.5: Irregular fluctuation of wind speed. (Provided by Prof. Tamagawa, Gifu University).

the right side that has the effect of dissipating the energy of fluid motion. This includes the second derivative with respect to the space variable, so as can be seen from the relation as "for $v = e^{i\boldsymbol{k}\cdot\boldsymbol{x}}$, $\nu\nabla^2 v = -\nu k^2 e^{i\boldsymbol{k}\cdot\boldsymbol{x}}$," it works more strongly for components with larger k, and hence shorter wavelengths.

Energy cascade

In a turbulent state, it is believed that the kinetic energy is first supplied to a relatively large scale (i.e., low wavenumber) of the size of the target object (for example, vehicle body) or the size of the container (for example, the dimension of the room). Then the energy is gradually passed to a smaller scales (i.e., higher wavenumbers) by the nonlinear effect, and finally dissipated in a very high wavenumber region where the viscosity works strongly. This flow of energy in the wavenumber space is called the **energy cascade**.

As a concrete example, let us consider the turbulent motion around a car of length l [m] and running at a speed v [m/s]. The typical time scale T of this turbulent motion can be estimated by $T = l/v$, and the turbulent kinetic energy per unit mass (1 kg) of air can be estimated by u^2. Then, the energy dissipation rate ε, that is, the amount of decrease of energy per second can be estimated by

$$\varepsilon \sim u^2/T \sim u^3/l. \tag{8.25}$$

If the car continues to run at speed v, the energy of turbulence around it should also be kept steady, so ε estimated by (8.25) is not only a measure of the amount of energy dissipation per unit time due to viscosity, but also a measure of the amount of energy supplied per unit time from the car to the turbulent motion to compensate for the dissipation. Thus, ε gives a measure of the amount of energy flow from the low wavenumber region to the high wavenumber region per second.

To how high a wavenumber does this energy cascade in wavenumber space last? The energy flow is represented by ε of (8.25), and the viscous effect by which the cascade is stopped is represented by the kinematic viscosity ν in the NS equation, the length scale l_ν for which the cascade stops should be determined from these two. Considering that the unit of ε is m^2/s^3 and the unit of v is m^2/s, l_v should be of the order of

$$l_v = \left(\frac{v^3}{\varepsilon}\right)^{1/4}. \tag{8.26}$$

For example, when a vehicle of length $l = 4$ m travels at speed of 60 km/h ($v = 16.7$ m/s) in the air ($v = 1.5 \times 10^{-5}$ m^2/s),

$$\varepsilon \sim 1.2 \times 10^3 \, m^2/s^3, \quad l_v \sim 4.1 \times 10^{-5} \, m = 0.041 \, mm. \tag{8.27}$$

This result shows that the energy of the vortex motion of several meters in size, which is directly produced by the running vehicle, is cascaded to smaller and smaller vortices up to extremely small motion of about 0.04 mm, where the kinetic energy is converted to heat by viscosity and disappears. The wavenumbers k_l, k_v corresponding to l, l_v can be introduced by $k_l = 1/l$, $k_v = 1/l_v$. From (8.25) and (8.26), the ratio of the length l of the large vortex which is supplied energy directly from the vehicle and the length l_v of the minimum-size vortex generated by the energy cascade is given by

$$\frac{l}{l_v} = \frac{k_v}{k_l} = \text{Re}^{3/4}, \quad \text{Re} = \frac{ul}{v}. \tag{8.28}$$

Here Re is the **Reynolds number**. Generally, the larger the Re, the more unstable the flow and the more likely it is to be turbulent. In the case of the turbulent flow around the running vehicle above, Re $\sim 4.5 \times 10^6$. According to (8.28), l_v should be about 1/100,000 of l, which is consistent with (8.27).

Kormogorov's $k^{-5/3}$ spectrum

Turbulence is ubiquitous and the causes of generation of turbulence vary widely. Some is generated by a running vehicle as above, some is generated by stirring a spoon in a coffee cup, some is generated by water flowing rapidly in the water pipe, and some is generated by the intense tidal current in the sea, and so on. Reflecting such individual situations, the shape of the energy spectrum of turbulence does not have universal properties that are common to all turbulence, at least with regard to the low wavenumber $k \approx k_l$ where the fluid motion is directly excited by external forces.

However, as the energy flows to higher wavenumbers by the cascade process, the memory of how the energy was supplied at $k \approx k_l$ is gradually lost. As a result, the energy spectrum in the region where the wavenumber is relatively high will have a universal property, regardless of how the turbulence is generated. Normally, the Reynolds number Re takes extremely large values in a turbulent state. In that case, according to (8.28), the wavenumber $k \sim k_l$ directly excited by external energy supply and the wavenumber $k \sim k_v$ where the cascade stops due to viscosity are very far apart. Therefore, an intermediate wavenumber region $k_l \ll k \ll k_v$ exists widely between the two. In this intermediate region, the memory of how the turbulence is generated is lost because $k \gg k_l$, and at the same time, k is not large enough for the effect of viscosity to be effective because $k \ll k_v$. An intermediate region with such properties is called the **inertial subrange**.

There are only two physical quantities that affect the spectral shape in the inertial subrange: the wavenumber k of interest and the energy flux ε that flows over there per unit time. We introduce the wavenumber spectrum $E(k)$ of turbulent motion by

$$\overline{\frac{1}{2}|\boldsymbol{v}|^2} = \int_0^\infty E(k)\,dk. \tag{8.29}$$

If the dimension of each related quantities is represented by [ ],

$$[\boldsymbol{v}] = LT^{-1}, \quad [k] = L^{-1}, \quad [E(k)] = L^3 T^{-2}, \quad [\varepsilon] = L^2 T^{-3}. \tag{8.30}$$

From these, it is inferred that $E(k)$ in the inertial subrange is in the form of

$$E(k) = C_K \varepsilon^{2/3} k^{-5/3} \tag{8.31}$$

only by dimensional consideration. This spectrum is called the **Kolmogorov's $k^{-5/3}$ spectrum**, and has been widely observed in experiments and in-situ observation with sufficiently large Re (see Fig. 8.6). C_K is a dimensionless universal constant called the Kolmogorov constant, which is known to take a value of about $C_K = 1.4 \sim 1.8$. (For general knowledge on turbulence in fluid mechanics, see, for example, [1, 13] and Chapter 13 of [8].)

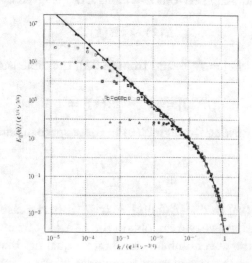

Figure 8.6: Kolmogorov's $k^{-5/3}$ spectrum (reprinted from [5] with permission).

8.4.2   POWER–LAW SPECTRUM OF OCEAN WAVES

Let us try to apply the argument for hydrodynamic turbulence as above to the situation where waves of various wavenumbers coexist and exchange energy with each other by nonlinear interactions like ocean waves. A state in which many waves coexist and interact with each other is called **wave turbulence** or **weak turbulence**. However, as we will see below, it is not as easy to derive a Kolmogorov-like power law spectrum for wave turbulence as in the case of fluid dynamic turbulence above.

Failure of argument based on dimension only

   In the case of an ocean wave field in which an infinite number of gravity waves coexist, the target spectrum for which the power law is sought is the wavenumber spectrum $E(k)$ which can be introduced by

$$\overline{\eta^2} = \iint E(\boldsymbol{k})\,d\boldsymbol{k} = \int E(k)\,dk. \tag{8.32}$$

Note that the difference between the wavenumber spectrum $E(k)$ and the vector wavenumber spectrum $E(\boldsymbol{k})$ is important. Since

$$\iint E(\boldsymbol{k})d\boldsymbol{k} = \int_0^\infty \left[\int_0^{2\pi} E(\boldsymbol{k})\,k\,d\theta \right] dk = \int_0^\infty E(k)dk,$$

$$\longrightarrow \quad E(k) = \int_0^{2\pi} E(\boldsymbol{k})\,k\,d\theta, \tag{8.33}$$

the rough magnitude relationship between $E(k)$ and $E(\boldsymbol{k})$ is given by

$$E(k) \sim k E(\boldsymbol{k}). \tag{8.34}$$

From $[\eta] = L$, $[k] = L^{-1}$, $[d\boldsymbol{k}] = L^{-2}$, and from (8.32), the dimensions of $E(k)$ and $E(\boldsymbol{k})$ are as follows:

$$[E(k)] = L^3, \quad [E(\boldsymbol{k})] = L^4. \tag{8.35}$$

Since the energy flux $\varepsilon(k)$ flowing through $E(k)$ in the k space satisfies

$$\frac{\partial E(k)}{\partial t} + \frac{\partial \varepsilon(k)}{\partial k} = 0, \tag{8.36}$$

the dimension of $\varepsilon(k)$ is $[\varepsilon(k)] = T^{-1}[E(k)][k] = L^2 T^{-1}$. In the case of the Kolmogorov spectrum of fluid dynamical turbulence, we can obtain the $k^{-5/3}$ spectrum immediately by combining the energy flux ε and the wavenumber k to obtain a quantity having the dimension of the spectrum $E(k)$. However, in the case of wave turbulence of ocean waves, since

$$[E(k)] = L^3, \quad [\varepsilon(k)] = L^2 T^{-1}, \quad [k] = L^{-1}, \tag{8.37}$$

if we try to make a quantity from ε and k, having the same dimension as $E(k)$, ε cannot enter, and it does not work. After all, it seems that we need to discuss beyond the level that relies solely on dimensions.[4]

Derivation of power law spectrum

Let us write τ a typical time scale of the spectral change caused by nonlinear effects. This is another time scale that is much longer than the linear time scale estimated by the reciprocal $1/\omega$ of the frequency. Evaluating the time derivative of $E(k)$ using this τ in (8.36),

$$\varepsilon(k) \sim \tau^{-1} E(k) k. \tag{8.38}$$

If there is a wavenumber region corresponding to the inertial subrange of fluid dynamic turbulece also for wave turbulence, then ε should stay constant ($= \varepsilon_0$) there. Therefore, if τ can be somehow estimated, we can estimate the form of $E(k)$ from this equation. And for this purpose we can use the energy balance equation (8.20).

In the energy balance equation (8.20), among the three source terms, leaving only S_{nl} that produces energy cascade that we are interested in, and assuming that the wave field is spatially uniform and $E(k)$ does not depend on location, the governing equation of $E(k)$ becomes simply

$$\frac{dE(k)}{dt} = S_{nl}. \tag{8.39}$$

Since the dimensions of both sides of this equation are equal, the dimension of the coupling coefficient W_{1234} in S_{nl} should be $L^{-4}T^{-2}$, so its magnitude can be estimated as $W \sim k^4 \omega^2$. If we estimate the magintude of both sides of (8.39) using the expression (8.21) for S_{nl}, $E(k) \sim E(k)/k$, and above estimate for W, we can get an evaluation for τ as follows[5]:

$$\tau^{-1} k^{-1} E(k) \sim \left(k^4 \omega^2\right) \left(\frac{E(k)}{k}\right)^3 \left(k^{-2}\right) \left(\omega^{-1}\right) \left(k^2\right)^3 = g^{1/2} k^{11/2} E^3$$
$$\longrightarrow \quad \tau^{-1} = g^{1/2} k^{13/2} E^2. \tag{8.40}$$

Substituting this into (8.38),

$$\varepsilon(k) \sim \left(g^{1/2} k^{13/2} E^2\right) E k = g^{1/2} k^{15/2} E^3. \tag{8.41}$$

---

[4]In fluid dynamic turbulence, the characteristic time scale at each wavenumber k is uniquely determined from the spectrum intensity $E(k)$ and k there. On the other hand, in wave turbulence, at each wavenumber k, there are two largely different time scales: one is the "linear time scale" determined from the frequency $\omega(k)$ and the other is the "nonlinear time scale" related to the temporal change of $E(k)$. This difference in situation is the reason why a simple dimensional analysis that has succeeded in fluid dynamic turbulence does not work in wave turbulence.

[5]Since $\delta(\boldsymbol{k})$ becomes 1 when integrated with respect to $\boldsymbol{k}$, its dimension is equal to $[\boldsymbol{k}]^{-1}$, and since $\delta(\omega)$ becomes 1 when integrated with respect to ω, its dimension is equal to $[\omega]^{-1}$.

From this, a power-law spectrum like

$$E(k) = \varepsilon_0^{1/3} g^{-1/6} k^{-5/2} \tag{8.42}$$

is obtained in the inertial subrange in which $\varepsilon(k) = \varepsilon_0$ holds. From (8.6), when the dispersion relation is $\omega^2 = gk$, the wavenumber spectrum $E(k)$ and the frequency spectrum $F(\omega)$ are related as $F(\omega) = E(k) \cdot 2\omega/g$. Therefore, $F(\omega)$ corresponding to (8.42) is

$$F(\omega) = \varepsilon_0^{1/3} g^{-1/6} \left(\frac{\omega^2}{g} \right)^{-5/2} \left(\frac{2\omega}{g} \right) = 2 \varepsilon_0^{1/3} g^{4/3} \omega^{-4}. \tag{8.43}$$

In this way, we can derive a frequency spectrum with the same power law as ω^{-4} often observed in developed ocean waves [6].

Thus, the power-law spectrum $F(\omega) \propto \omega^{-4}$ of ocean surface waves does not contradict with (the counterpart of) the Kolmogorov spectrum that is expected when the concept of inertial subrange of fluid dynamic turbulence is directly applied also to the ocean wave field. However, in the case of ocean waves, the spectrum width in the wavenumber and frequency spaces is not so wide as compared to fluid dynmaic turbulence, and it is difficult to be in a situation where the power law can be clearly detected as in the case of fluid dynamic turbulence shown in Fig. 8.6. Also, in the above analysis, we have ignored all other source terms such as the wind effect S_{in} and the wave breaking effect S_{ds} and have derived $F(\omega) \propto \omega^{-4}$ based on the premise that only the nonlinearity effect S_{nl} works. However, in the real ocean, it is also reported that the wavenumber region where these three source terms are dominant is not so clearly separated. Therefore, it is not obvious from the beginning whether there is a wavenumber region to which the concept of inertial subrange can be applied. In this sense, there may still be room for consideration in immediately understand the frequency spectrum of the power law $F(\omega) \propto \omega^{-4}$ observed in the actual ocean wave field as having the same origin as the Kolmogorov spectrum of fluid dynamic turbulence. Not only the problem of the power-law spectrum as above, there still remain many interesting problems to be solved or clarified in the reseach field of wave turbulence. (For example, see [10, 11], etc.)

Since the SMB method developed during World War II, research on wave prediction has made great strides, and today, anyone can know the forecast for the significant wave height, significant wave period and wave directions in all sea area of the world from the Internet. Figure 8.7 is downloaded from one of such sites, `http://polar.ncep.noaa.gov/waves/`. The site displays the wave forecast obtained using a program package for numerical wave prediction called WAVEWATCH III developed by the National Center for Environmental Prediction (NCEP) of the National Ocean and Atmosphere Administration (NOAA). This web page shows the results of forecast for more than a week (180 h) by an animation, so you can see how an area of large wave height generated by a storm travels through the ocean.

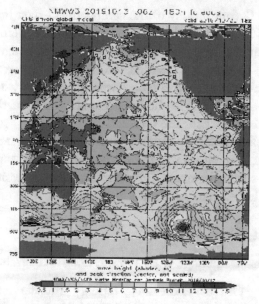

Figure 8.7: Numerical wave forecast (from `http://polar.ncep.noaa.gov/waves/`).

COFFEE BREAK: WAVE FORECAST METHOD DEVELOPED FOR WAR

The averaging procedure that defines the significant wave height is not a mean in the ordinary sense which is taken over all the individual waves, but it employs a somewhat strange quantity like the mean over only the largest one third. This based on the research of ocean wave forecasting methods conducted by the research team led by two oceanographers H. U. Sverdrup and W. H. Munk in the United States during World War II in connection with the military operations.

The human eye seems to tend to focus on the larger waves rather than the smaller ones, and Sverdrup and Munk found that the mean wave height over the largest one third agrees well with the wave height estimated by the well-trained observer looking at the sea. From this, the significant wave height $H_{1/3}$ began to be used as a representative value of the wave height. They first of all introduced clearly defined quantities of $H_{1/3}$, $T_{1/3}$ as representative quantities to be predicted for the field of ocean wave prediction, for which any quantitative method of expression had not yet been established. Then, by theoretical analysis and organizing observation data that were accumulated so far in a unified manner, they constructed a framework to predict $H_{1/3}$ and $T_{1/3}$ as functions of wind speed, fetch (distance to the upwind coastline) and duration (the time elapsed since the wind started to blow). This wave forecasting method is called the **SMB method** after Sverdrup, Munk, and Bretschneider who later improved it.

The Normandy landing operation by the Allied Forces is well known as an operation that greatly influenced the outcome of World War II in Europe. It is said that the SMB method is also related to its success. Before the day of the campaign (D-day: June 6, 1944), the sea near the Dover Strait was heavily roughened by a strong storm. The Germans defending the coast of Normandy believed that the storm would not settle for several days and there would be no enemy landing. However, by using the weather forecast and the wave prediction method, the Allied Forces more accurately predicted the date when the storm would weaken and the date when the wave height would decrease to the level that landing operation becomes possible. As a result, it became a surprise attack for the Germans.

Although it is sad, electric and electronic technologies such as radar, transportation means such as aircraft and rockets, and nuclear research..... no matter what you think, it is an undeniable historical fact that various studies have made great strides out of necessity every time a war occurs that many human beings kill each other for life.

8.5   REFERENCES

[1] U. Frisch. *Turbulence*. Cambridge U.P., 1995. DOI: 10.1017/cbo9781139170666 173

[2] K. Hasselmann.   On the non-linear energy transfer in a gravity-wave spectrum. part 1. general theory. *Journal of Fluid Mechanics*, 12:481–500, 1962. DOI: 10.1017/s0022112062000373 169

[3] L. H. Holthuijsen. *Waves in Oceanic and Coastal Waters*. Cambridge U.P., 2007. DOI: 10.1017/cbo9780511618536 164

[4] P. Janssen. *The Interaction of Ocean Waves and Wind*. Cambridge U.P., 2004. DOI: 10.1017/cbo9780511525018 168

[5] S. Kida and S. Yanase. *Ranryu Rikigaku (Turbulence Mechanics)*. Asakura Publishing, 1999. (in Japanese). 173

[6] S. A. Kitaigorodskii. On the theory of the equilibrium range in the spectrum of wind-generated gravity waves. *Journal of Physical Oceanography*, 13:816–827, 1983. DOI: 10.1175/1520-0485(1983)013<0816:ottote>2.0.co;2 176

[7] G. J. Komen, L. Cavaleri, K. Hasselmann, S. Hasselmann, and P. A. E. M. Janssen. *Dynamics and Modelling of Ocean Waves*. Cambridge U.P., 1994. DOI: 10.1017/cbo9780511628955 168

[8] P. K. Kundu and I. M. Cohen. *Fluid Mechanics*, 3rd ed., Elsevier, 2004. 173

[9] S. R. Massel. *Ocean Surface Waves—Their Physics and Prediction*. World Scientific, 1996. DOI: 10.1142/9789812795908 165

[10] S. Nazarenko. *Wave Turbulence*. Springer, 2011. DOI: 10.1007/978-3-642-15942-8 170, 176

[11] A. C. Newell and B. Rumpf. Wave turbulence: A story far from over. In V. Shrira and S. Nazarenko, Eds., *Advances in Wave Turbulence*. World Scientific, 2013. DOI: 10.1142/9789814366946_0001 176

[12] O. M. Phillips. *The Dynamics of the Upper Ocean*. Cambridge U.P., 1966. DOI: 10.2307/3614226 168

[13] H. Tennekes and J. L. Lumley. *A First Course of Turbulence*. The MIT Press, 1972. 173

[14] Y. Yoshimi Goda. *Random Seas and Design of Maritime Structures*. World Scientific, 2010. DOI: 10.1142/7425 165

[15] V. E. Zakharov, V. S. L'vov, and G. Falkovich. *Kolmogorov Speacta of Turbulence I—Wave Turbulence*. Springer, 1992. DOI: 10.1007/978-3-642-50052-7 170

APPENDIX A

Conservation Law in 3D

In Chapter 1, we discussed the conservation law assuming that the space is 1D. This appendix introduces how the flux is defined and how the conservation law is modified when the space becomes 3D.

A.1 FLUX DENSITY VECTOR

As discussed in Section 2.2, the driving force that changes the temperature distribution of wire in time is the heat flux, that is, the flow of heat from a place with high temperature to the place with low temperature. Not only the case of temperature of a wire, generally the flux of a physical quantity is essentially important when considering how the spatial distribution of the physical quantity changes in time. If the space is 1D like the heat flow in the wire, the heat flux is defined as the amount of heat that passes through the point of interest x per second, so the unit of heat flux is Joule/second. However, the flux in a continuous medium spread in a 3D space is not that simple.

The flux of a certain physical quantity P that takes a scalar value at a point x in a 3D space can be expressed by indicating the intensity and the direction of the flow. The flux at a certain point can be expressed by a single vector by linking the intensity and the direction of the flux to the magnitude and the direction of the vector, respectively. Then, with what kind of quantity should we express the "intensity of flux" concretely? In the case of heat flux in the wire, it was only necessary to consider the amount of heat (Joule) passing through the point x per second. If we try to think about the same thing at a certain point in 3D space, we assume a plane perpendicular to the direction of heat flow at that point, opening a small window there, and measuring the amount of heat passing through that small window. However, since the amount of heat passing through the window depends on the area of the window, the area of the window must be normalized. Therefore, in order to rationally express the intensity of the flow of the physical quantity P at a point in 3D space, the amount of P that passes through a window of unit area perpendicular to the flow direction per unit time is used. A vector that has the "flow intensity" of P thus defined as the magnitude and has the direction of the flow of P as the direction is called the **flux density vector** of P. When the unit of P is denoted as $\odot$, the unit of the flux density vector is $\odot /m^2$ s.

A.2   CONSERVATION LAW IN INTEGRAL FORM

Take an arbitrary 3D region V fixed in space, and consider the conservation law of a certain physical quantity P there. Let the unit of P be $\odot$, and let the density and the flux density of P be $\rho(\boldsymbol{x}, t)$ [$\odot/\text{m}^3$] and $\boldsymbol{q}(\boldsymbol{x}, t)$[$\odot/\text{m}^2$ s], respectively. Since there is an inflow and outflow of P through the surface S of V, the total amount of P in V is not constant in time but will constantly change. The conservation law of P requires that the increment of P in V that occurs during a certain time is equal to the amount of net inflow through the surface S during that time. By calculating separately the increment of P in V and the net inflow of P through S and equating them, we can derive a formula that expresses the conservation law of P as follows.

The total amount of P in V at time t is given by the volume integral

$$M(t) = \iiint_V \rho(\boldsymbol{x}, t)\, dV, \tag{A.1}$$

so its increment per unit time is given by[1]

$$\frac{dM(t)}{dt} = \iiint_V \frac{\partial \rho(\boldsymbol{x}, t)}{\partial t}\, dV. \tag{A.2}$$

On the other hand, when the flux density of P is $\boldsymbol{q}(\boldsymbol{x}, t)$, the amount of P flowing out through the surface element dS per unit time is $\boldsymbol{q} \cdot \boldsymbol{n}\, dS$, where $\boldsymbol{n}$ being the outward unit normal vector of dS. Therefore, the total amount of P flowing into V through the entire surface S per unit time is given by the surface integral

$$- \iint_S \boldsymbol{q} \cdot \boldsymbol{n}\, dS. \tag{A.3}$$

By equating (A.2) and (A.3), we obtain the conservation law of P in the integral form as follows:

$$\iiint_V \frac{\partial \rho}{\partial t}\, dV + \iint_S \boldsymbol{q} \cdot \boldsymbol{n}\, dS = 0. \tag{A.4}$$

This is the 3D counterpart of the 1D conservation law (1.30).

A.3   CONSERVATION LAW OF DIFFERENTIAL FORM

The following **Gauss's divergence theorem** is well known in vector analysis. That is, let V be an arbitrary bounded closed region in 3D space, S be the boundary (i.e., the surface) of V, $\boldsymbol{n}$ be the outward unit normal vector of S, and $\boldsymbol{u}(\boldsymbol{x})$ be a smooth vector field defined in V, then the divergence theorem teaches us that

$$\iiint_V \operatorname{div} \boldsymbol{u}\, dV = \iint_S \boldsymbol{u} \cdot \boldsymbol{n}\, dS. \tag{A.5}$$

---

[1]Since the integration region V does not change with time in this case, the time derivative can simply be put into the integral.

Here, div $\boldsymbol{u}$ is a scalar field defined for a vector field

$$\boldsymbol{u}(x) = u_1(x, y, z)\boldsymbol{i} + u_2(x, y, z)\boldsymbol{j} + u_3(x, y, z)\boldsymbol{k} \tag{A.6}$$

by

$$\operatorname{div} \boldsymbol{u} \equiv \frac{\partial u_1}{\partial x} + \frac{\partial u_2}{\partial y} + \frac{\partial u_3}{\partial z}, \tag{A.7}$$

and is called the **divergence** of $\boldsymbol{u}$. If we introduce the differential operator called **nabla** by

$$\nabla \equiv \frac{\partial}{\partial x}\boldsymbol{i} + \frac{\partial}{\partial y}\boldsymbol{j} + \frac{\partial}{\partial z}\boldsymbol{k}, \tag{A.8}$$

div $\boldsymbol{u}$ can also be written as $\nabla \cdot \boldsymbol{u}$.

By converting the surface integral of the conservation law of integral form (A.4) into a volume integral using the divergence theorem and combining the two volume integrals into one, we obtain

$$\iiint_V \left(\frac{\partial \rho}{\partial t} + \operatorname{div} \boldsymbol{q} \right) dV = 0. \tag{A.9}$$

In order for this to hold for any volume region V, the integrand must be 0 at any point in V, which yields the following conservation law of P in differential form:

$$\frac{\partial \rho}{\partial t} + \operatorname{div} \boldsymbol{q} = 0. \tag{A.10}$$

This is the 3D counterpart of the 1D conservation law in the differential form (1.10).

APPENDIX B

System of Simultaneous Wave Equations

For example, when sound wave is transmitted, the pressure, density, flow velocity, etc. of the air change simultaneously. In this appendix, we will consider how to handle this type of wave that simultaneously conveys changes in multiple physical quantities. We also explain that this type of wave can be reduced to a single wave equation as described in Chapter 1 under the situation called "simple wave."

B.1 HYPERBOLIC EQUATION

In Chapter 1, we treated the wave equation $u_t + c(u) u_x = 0$ for a single physical quantity $u(x, t)$. The natural extension of this to the case where changes in multiple physical quantities are transmitted simultaneously with a wave would be

$$\frac{\partial \boldsymbol{u}}{\partial t} + A(\boldsymbol{u})\frac{\partial \boldsymbol{u}}{\partial x} = \boldsymbol{0}, \tag{B.1}$$

where $\boldsymbol{u}(x,t) = {}^t(u_1(x,t), \ldots, u_n(x,t))$ is a column vector consisting of n dependent variables (${}^t\boldsymbol{u}$ denotes the transpose of $\boldsymbol{u}$), and $A(\boldsymbol{u})$ is an $n \times n$ square matrix. $A(\boldsymbol{u})$ generally depends on $\boldsymbol{u}$, but does not include the derivative of $\boldsymbol{u}$. (B.1) is linear in $\boldsymbol{u}$ when A does not depend on $\boldsymbol{u}$, but becomes nonlinear when it depends on $\boldsymbol{u}$.[1] For example, if there are two dependent variables, the concrete expression of (B.1) becomes

$$\frac{\partial u_1}{\partial t} + a_{11}(u_1, u_2)\frac{\partial u_1}{\partial x} + a_{12}(u_1, u_2)\frac{\partial u_2}{\partial x} = 0, \tag{B.2a}$$

$$\frac{\partial u_2}{\partial t} + a_{21}(u_1, u_2)\frac{\partial u_1}{\partial x} + a_{22}(u_1, u_2)\frac{\partial u_2}{\partial x} = 0. \tag{B.2b}$$

In Section 1.2.2, when we obtained an Equation (1.10) including two quantities, the density ρ and the flux q, from the conservation law, we mentioned that there are two ways as follows to make the problem "closed," that is, to convert the problem into the situation where the number of equations and the number of unknowns are equal, (i) deriving a new equation

---

[1](B.1) is nonlinear when A depends on $\boldsymbol{u}$, but it is linear with respect to the derivative $\boldsymbol{u}_t$, $\boldsymbol{u}_x$. A partial differential equation that is linear with respect to the highest-order derivative like (B.1) is called a "quasi-linear PDE."

for $\partial q / \partial t$ from another conservation law or some physical laws, or (ii) assuming an equation of state like $q = q(\rho)$. In Section 1.2, the method of (ii) is adopted and a single wave equation (1.16) directly obtained from (1.10), but when the method (i) is adopted here we often end up with a system of simultaneous equations such as (B.1).

In the water wave problem, the surface displacement $\eta(x, t)$ and the horizontal flow velocity $u(x, t)$ are governed by the **long wave equation**,[2]

$$\frac{\partial \eta}{\partial t} + \frac{\partial[(h + \eta)u]}{\partial x} = 0, \qquad \frac{\partial u}{\partial t} + u \frac{\partial u}{\partial x} + g \frac{\partial \eta}{\partial x} = 0, \tag{B.3}$$

when the wavelength is much longer than the water depth h. This system of equations can be cast into the form of (B.1) by putting

$$\boldsymbol{u} = \begin{pmatrix} \eta \\ u \end{pmatrix}, \qquad A = \begin{pmatrix} u & h + \eta \\ g & u \end{pmatrix}. \tag{B.4}$$

In the motion of ideal gas, if there is no dissipation such as viscosity or heat conduction and hence the entropy can be assumed to be constant, the basic equations are given by

$$\frac{\partial \rho}{\partial t} + u \frac{\partial \rho}{\partial x} + \rho \frac{\partial u}{\partial x} = 0, \qquad \frac{\partial u}{\partial t} + u \frac{\partial u}{\partial x} + \frac{c^2}{\rho} \frac{\partial \rho}{\partial x} = 0, \tag{B.5}$$

where ρ is density, u is flow velocity, c is sound speed given by $c^2 = \gamma p / \rho$ with p is pressure, γ is the ratio c_p / c_v of specific heat at constant pressure c_p to that at constant volume c_v, and $\gamma \approx 1.4$ in the air. This can also be cast into the form (B.1) by putting

$$\boldsymbol{u} = \begin{pmatrix} \rho \\ u \end{pmatrix}, \qquad A = \begin{pmatrix} u & \rho \\ c^2/\rho & u. \end{pmatrix}. \tag{B.6}$$

In (B.1), when the n eigenvalues $\lambda^{(1)}, \ldots, \lambda^{(n)}$ of A are all real and the corresponding eigenvectors are linearly independent, (B.1) is called **hyperbolic** quasi-linear PDEs. And then, the curve $C^{(i)}$ on the x-t plane given by $\frac{dx}{dt} = \lambda^{(i)}$ ($i = 1, 2, \ldots, n$) is called the **(nth) characteristic curve**.[3] Since $\frac{dx}{dt}$ expresses velocity, so the definition of characteristic curve $\frac{dx}{dt} = \lambda$ means that the eigenvalue λ of A corresponds to the velocity at which the characteristic curve is transmitted. In the case of long wave equation (B.3), the two eigenvalues of A are given by

$$\lambda_{\pm} = u \pm \sqrt{g(h + \eta)} \tag{B.7}$$

which are both real and different with each other, so the corresponding two eigenvalues are always linearly independent, so this system is a hyperbolic quasi-linear PDSs. Similarly, in the case of the ideal gas equation (B.5), the eigenvalues of A are given by two different real numbers $\lambda_{\pm} = u \pm c$, so this system is also hyperbolic.

---

[2]For the derivation of this equation, see Appendix F.

[3]The characteristic curve defined in this way matches the characteristic curve introduced in Chapter 1 when $n = 1$.

B.2   MECHANISM OF TEMPORAL EVOLUTION OF HYPERBOLIC SYSTEM

Let $l^{(i)}$ be the left eigenvector[4] corresponding to the eigenvalue $\lambda^{(i)}$ of A and $x = X^{(i)}(t)$ be the parametric representation of the corresponding characteristic curve $C^{(i)} : dx/dt = \lambda^{(i)}$.

Taking the product of (B.1) and $l^{(i)}$, and using $l^{(i)}A = \lambda^{(i)}l^{(i)}$, we obtain

$$l^{(i)} \left(\frac{\partial u}{\partial t} + A \frac{\partial u}{\partial x} \right) = l^{(i)} \left(\frac{\partial}{\partial t} + \lambda^{(i)} \frac{\partial}{\partial x} \right) u = 0. \tag{B.8}$$

In the same way as (1.18), the time derivative d/dt along the characteristic curve $C^{(i)}$ is given by

$$\frac{d}{dt} = \frac{\partial}{\partial t} + \frac{dX^{(i)}(t)}{dt} \frac{\partial}{\partial x} = \frac{\partial}{\partial t} + \lambda^{(i)} \frac{\partial}{\partial x}, \tag{B.9}$$

therefore (B.8) can be written as

$$l^{(i)} \frac{du}{dt} = 0 \quad \text{along } C^{(i)}. \tag{B.10}$$

This ordinary differential equation is called **the characteristic form**. If we write $l^{(i)}$ as $l^{(i)} = (l_1^{(i)}, \ldots, l_n^{(i)})$ and the small change du occurring in short time dt as $du = (du_1, \ldots, du_n)$, then (B.10) can be written as

$$l_1^{(i)} du_1 + l_2^{(i)} du_2 + \cdots + l_n^{(i)} du_n = 0 \quad \text{along } C^{(i)}. \tag{B.11}$$

Thus, a characteristic form defines a linear relationship between the changes of n dependent variables along $C^{(i)}$. Since the coefficient $(l_1^{(i)}, \ldots, l_n^{(i)})$ are different for each characteristic curve, the different characteristic curves require different linear relationships among du_i.

As discussed in Chapter 1, in the case of a single wave equation

$$\frac{\partial u}{\partial t} + c(u) \frac{\partial u}{\partial x} = 0, \tag{B.12}$$

there is only one kind of characteristic curves, and the waveform at an arbitrary time is determined by a constant value of the dependent variable u which is carried along the characteristic curves. Then how does the solution $u(x, t)$ be determined for a system of hyperbolic equations with multiple dependent variables? To understand the mechanism, assume that we know $u(x, t_0)$ at all point x at some time t_0, and let's think about how u at point P a short time Δt later is determined by looking at Fig. B.1. For simplicity, we assume that $n = 2$ here. At time t_0, two characteristic curves are radiated from each x, of which two pass through the point P at $t_0 + \Delta t$. The characteristic forms (B.10) for these two give conditions for the increment du, which determine u at P as follows.

---

[4]A non-zero vector l that satisfies $lA = \lambda l$ is called the left eigenvector of A. Taking the transpose of both sides yields ${}^tA\,{}^tl = \lambda\,{}^tl$. This shows that if we find the ordinary eigenvector (right eigenvector) for the transpose of A first and transpose it, then we can get the left eigenvector of A to the same eigenvalue.

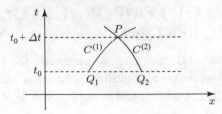

Figure B.1: Mechanism of determining the solution of hyperbolic system.

Let Q_1 and Q_2 be the positions of two characteristic curves $C^{(1)}$ and $C^{(2)}$ that reach point P at time $t_0 + \Delta t$ at time t_0, respectively. If Δt is sufficiently small and du/dt of (B.10) can be evaluated by a forward difference, the characteristic forms along $C^{(1)}$ and $C^{(2)}$ can be written, respectively, as

$$\boldsymbol{l}^{(1)}(Q_1)\{\boldsymbol{u}(P) - \boldsymbol{u}(Q_1)\} = 0, \tag{B.13a}$$

$$\boldsymbol{l}^{(2)}(Q_2)\{\boldsymbol{u}(P) - \boldsymbol{u}(Q_2)\} = 0. \tag{B.13b}$$

Here, the unknown quantities are only $\boldsymbol{u}(P)$, and the rest are all known quantities that are evaluated at time t_0. (B.13) gives the system of linear equations $L\,\boldsymbol{u}(P) = \boldsymbol{c}$ for $\boldsymbol{u}(P)$, where L is a 2×2 matrix with $\boldsymbol{l}^{(i)}$ $(i = 1, 2)$ as the ith row vector, and $\boldsymbol{c}$ is a column vector with $\boldsymbol{l}^{(i)}(Q_i)\boldsymbol{u}(Q_i)$ $(i = 1, 2)$ as the ith component. The linear independence of the eigenvectors included in the definition of "hyperbolicity" guarantees that the coefficient matrix L is non-singular, so $\boldsymbol{u}(P)$ is uniquely determined. By repeating this procedure in time, $\boldsymbol{u}(x, t)$ at an arbitrary time is determined when the initial condition $\boldsymbol{u}(x, 0)$ is specified. In the above, we treated the case of $n = 2$ for simplicity. However, there is no essential difference in the process in which $\boldsymbol{u}(x, t)$ is determined in the case of general n. In this case, n different kinds of characteristic curves $C^{(i)}$ requiring n different linear relationships for the increment $d\boldsymbol{u}$ propagate at each characteristic velocity $\lambda^{(i)}$, and their intersection at one point P:(x, t) determines the value of $\boldsymbol{u}(x, t)$ there.

From the process of temporal evolution described above, it can be seen that the solution $\boldsymbol{u}(P)$ at the point P depends only on the triangular part between the fastest C^+ and the slowest C^- among the n characteristic curves that reach P (see Fig. B.2). This region is called the **domain of dependence** of P. Even if a point in the xt-plane is a point in the past viewed from P, if it is outside the domain of dependence, no matter how we change $\boldsymbol{u}$ at that point, the value of $\boldsymbol{u}$ at P is not affected. On the other hand, the wedge-shaped region between C^+ and C^- emanating from P as shown in Fig. B.2 is the region which is affected by the event at P, and is called the **range of influence** of P.

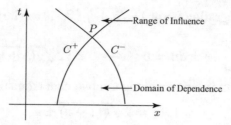

Figure B.2: Domain of dependence and range of influence.

B.3   RIEMANN INVARIANT

If (B.1) is linear, that is, A is a constant matrix independent of $\boldsymbol{u}$, its left eigenvector $\boldsymbol{l}^{(i)}$ is also a constant vector. Then, if we introduce $R^{(i)}$ by $R^{(i)} = \boldsymbol{l}^{(i)}\boldsymbol{u}$, (B.10) becomes

$$\frac{dR^{(i)}}{dt} = 0 \quad \text{along} \quad C^{(i)} : \frac{dX^{(i)}}{dt} = \lambda^{(i)}, \tag{B.14}$$

and $R^{(i)}$ is kept constant along the characteristic curve $C^{(i)}$.

On the other hand, in the nonlinear case, A, and hence $\boldsymbol{l}^{(i)}$ depends on $\boldsymbol{u}$ and changes with $\boldsymbol{u}$. In the case of $n \geq 3$, there is generally no integrating factor that converts $\boldsymbol{l}^{(i)}d\boldsymbol{u}$ to a total derivative of a certain quantity, and therefore it is not possible to rewrite (B.10) to the form (B.14). However, the existence of such an integrating factor is always guaranteed when $n = 2$ [1], so there always exist $R^{(i)}$ such that

$$\frac{dR^{(i)}}{dt} = 0 \quad \text{along} \quad C^{(i)} : \frac{dX^{(i)}}{dt} = \lambda^{(i)}, \quad (i = 1, 2). \tag{B.15}$$

Such a function of $\boldsymbol{u}$ that takes a constant value along the characteristic curve is called the **Riemann invariant**.

EXAMPLE 1: RIEMANN INVARIANT FOR LONG WAVE AND IDEAL GAS

1. Find the Riemann invariants for the long wave equation (B.3).

2. Find the Riemann invariants for the ideal gas equation (B.5).

[Answer]

1. In the case of long wave equations (B.3) and (B.4),

$$\boldsymbol{u} = \begin{pmatrix} \eta \\ u \end{pmatrix}, \quad \lambda_{\pm} = u \pm \sqrt{g(h + \eta)}, \quad \boldsymbol{l}_{\pm} = (\sqrt{g}, \pm\sqrt{h + \eta}), \tag{B.16}$$

and the characteristic form is given by

$$\sqrt{g}\,d\eta \pm \sqrt{h + \eta}\,du = 0 \quad \text{along} \quad C_{\pm} : \frac{dx}{dt} = u \pm \sqrt{g(h + \eta)}. \tag{B.17}$$

By multiplying the integrating factor $1/\sqrt{h + \eta}$, (B.17) can be written in a total differential form as

$$\sqrt{\frac{g}{h + \eta}}\, d\eta \pm du = 0 \quad \longrightarrow \quad d\left(2\sqrt{g(h + \eta)} \pm u\right) = 0. \tag{B.18}$$

From this, the Riemann invariants along the two characteristic curves $C_\pm$ are given by

$$R_\pm = 2\sqrt{g(h + \eta)} \pm u. \tag{B.19}$$

2. In the case of the ideal gas equations (B.5) and (B.6), $\boldsymbol{u}$, the eigenvalues, and the left eigenvectors of A are given by

$$\boldsymbol{u} = \begin{pmatrix} \rho \\ u \end{pmatrix}, \quad \lambda_\pm = u \pm c, \quad \boldsymbol{l}_\pm = (c, \pm\rho). \tag{B.20}$$

Therefore, the characteristic form becomes

$$c\, d\rho \pm \rho\, du = 0 \quad \text{along} \quad C_\pm : \frac{dx}{dt} = u \pm c. \tag{B.21}$$

These two types of waves ($\pm$) represent sound waves propagating in the positive and negative x directions. In an ideal gas with constant entropy, the pressure is expressed as a function of only ρ as $p = a\rho^\gamma$, where a is a constant determined by the entropy and γ is the adiabatic constant (i.e., ratio of the specific heats). In this case, the sound speed c is also a function only of ρ given by

$$c^2 = \frac{dp}{d\rho} = a\gamma\rho^{\gamma-1} \left(= \frac{\gamma p}{\rho}\right). \tag{B.22}$$

Dividing both sides of (B.21) by ρ and introducing a function $r(\rho)$ by

$$r(\rho) \equiv \int^\rho \frac{c(\rho)}{\rho}\, d\rho, \tag{B.23}$$

the characteristic form (B.21) becomes

$$dr \pm du = 0 \quad \longrightarrow \quad d(r \pm u) = 0, \tag{B.24}$$

indicating that the quantity defined by $R_\pm = r \pm u$ is constant along $C_\pm$. By differentiating (B.22),

$$2c\, dc = \frac{(\gamma - 1)c^2}{\rho}\, d\rho \quad \longrightarrow \quad r = \int \frac{c}{\rho}\, d\rho = \int \frac{2}{\gamma - 1}\, dc = \frac{2c(\rho)}{\gamma - 1}. \tag{B.25}$$

From this the Riemann invariant $R_\pm$ along $C_\pm$ can be written as

$$R_\pm = r \pm u = \frac{2c(\rho)}{\gamma - 1} \pm u. \tag{B.26}$$

♣

B.4   SIMPLE WAVE

As discussed above, in hyperbolic waves, there are n kinds of characteristic curves that carry the different linear relationships $\boldsymbol{l}^{(i)} d\boldsymbol{u} = 0$ $(i = 1, \ldots, n)$ between the changes of n dependent variables at the speed $\lambda^{(i)}$. And the intersection of these characteristic curves determines $\boldsymbol{u}(x, t)$.

When the initial disturbance is spatially localized, the waves (signals) emitted from it is gradually divided into n groups over time according to the speed of the characteristic curves (see Fig. B.3). If the fastest group of characteristic curves is written as C^+, the observer in the far right x_R of the initial disturbance will remain quiet without disturbance until the time when C_A^+ arrives. For the first non-zero $\boldsymbol{u}$ that is observed at x_R, only C^+ out of the n characteristic curves which intersect there to determine $\boldsymbol{u}$ there emanates from the disturbance region, while the remaining $(n-1)$ of them all emanate from the undisturbed uniform state. In this sense, for the observer at x_R, the first signal from the initial disturbance will be conveyed exclusively by C^+.

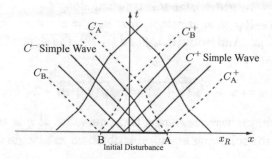

Figure B.3: Propagation of localized initial disturbance ($n = 2$).

The region and also the type of wave that is observed in the xt-space for which, of the n characteristic curves that determine $\boldsymbol{u}(x, t)$, $(n-1)$ are all emanated from the undisturbed state and only $C^{(i)}$ carries information of the initial disturbance as above is called the **simple wave (of the ith mode)**. It is mathematically proved that the next to the undisturbed region is always a simple wave region.

In addition, although not described in detail here, in the simple wave region of ith mode, the fact that all characteristic curves other than $C^{(i)}$ originate from the undisturbed state gives $(n-1)$ algebraic functional relationships between n dependent variables u_j $(j = 1, \ldots, n)$, and by using them we can represent all other variables algebraically with just one variable, for example u_1. From this, in the simple wave region, it is possible to reduce the fundamental system of Eq. (B.1) to a single wave equation of the form treated in Chapter 1 regardless of the number of dependent variables n. In the following example, let's confirm this by taking the long wave equation as an example.

EXAMPLE 2: SIMPLE WAVE FOR LONG WAVE EQUATION

For the long wave equation (B.3), find a single wave equation containing only $\eta(x, t)$ that holds in the simple wave region corresponding to the C_+ mode.

[Answer]

Since all C_- passing through this region start from the undisturbed region, the Riemann invariant R_- they carry is a constant value determined by the undisturbed state (i.e., $\eta = 0$, $u = 0$). Therefore, in this region,

$$R_- = 2\sqrt{g(h + \eta)} - u = 2\sqrt{gh} \ (=\text{constant}), \tag{B.27}$$

and u can be expressed by η as

$$u = 2\sqrt{g(h + \eta)} - 2\sqrt{gh}. \tag{B.28}$$

On the other hand, $R_+ = 2\sqrt{g(h + \eta)} + u$ is constant along C_+. Substituting (B.28) into this gives $R_+ = 4\sqrt{g(h + \eta)} - 2\sqrt{gh} = \text{constant}$, that is, $\eta = \text{constant}$, and so is u also constant from (B.28) along each C_+.[5] And from (B.16) and (B.28), the slope of C_+ (i.e., the wave propagation speed) λ_+ is given by

$$\lambda_+ = u + \sqrt{g(h + \eta)} = 3\sqrt{g(h + \eta)} - 2\sqrt{gh}. \tag{B.29}$$

As a result, in the C_+ simple wave region, a constant value of η is transmitted at the speed λ_+ given by (B.29), and the following equation is obtained as a single nonlinear wave equation expressing this fact:

$$\frac{\partial \eta}{\partial t} + c(\eta)\frac{\partial \eta}{\partial x} = 0, \qquad c(\eta) = 3\sqrt{g(h + \eta)} - 2\sqrt{gh}. \tag{B.30}$$

Of course, the corresponding wave equation for u can also be derived.

♣

. . . .

When a long wave such as a tsunami generated far away is transmitted to an undisturbed region with a depth h, the water surface displacement $\eta(x, t)$ is considered to follow (B.30). The propagation speed of tsunami is often said to be $\sqrt{gh}$, but a more accurate evaluation that takes into account nonlinear effects is $c(\eta) = 3\sqrt{g(h + \eta)} - 2\sqrt{gh}$ as given by (B.30). This converges to $\sqrt{gh}$ at the limit of $\eta/h \to 0$. The wave propagation velocity $c(\eta)$ is an increasing function of η, so the higher the water surface, the faster it propagates. As a result, as learned in Chapter 1,

---

[5]It should be noted that the meaning of "constant" differs between "R_- is constant" and "R_+ is constant." In the C_+ simple wave region, all C_- are emanated from undisturbed state, so the value of R_- carried by C_- are all equal. Therefore, R_- takes a constant value throughout this simple wave region regardless of (x, t). In contrast, R_+ is constant along each C_+, but different C_+ carry different values of R_+.

if there is a part of $\partial \eta / \partial x < 0$ in the initial waveform $\eta_0(x)$, the slope of that part gradually becomes steeper with time, and the slope diverges infinitely within a finite time. On the shallow coast, we can see the swells that are approaching the shore gradually lean forward and break. The nonlinearity also contributes to such a familiar phenomenon.

For more detailed information on the overall contents of this Appendix, refer to [2, 3], for example.

B.5   REFERENCES

[1] E. L. Ince. *Ordinary Differential Equations.* Dover, 1956. DOI: 10.2307/3605524 189

[2] T. Taniuti and K. Nishihara. *Nonlinear Waves.* Pitman, 1983. 193

[3] G. B. Whitham. *Linear and Nonlinear Waves.* John Wiley & Sons, 1974. DOI: 10.1002/9781118032954 193

APPENDIX C

Summary of Fourier Analysis

In analyzing wave phenomena, Fourier analysis is an indispensable tool. Here we summarize the minimum knowledge about it. The solution of the diffusion equation used in Chapter 2 is also given at the end of this Appendix as an application of Fourier analysis.

C.1   FOURIER SERIES

When $f(t)$ is a periodic function of period T and $f(t)$, and $f'(t)$ are both continuous, it is known that $f(t)$ can be expressed as

$$f(t) = \frac{1}{2}a_0 + \sum_{k=1}^{\infty} (a_k \cos \omega_k t + b_k \sin \omega_k t), \quad \omega_k = (2\pi/T)k \quad (k = 1, 2, \ldots), \quad (C.1)$$

by using appropriate coefficients a_k and b_k. The $\cos \omega_k t$ and $\sin \omega_k t$ appearing here represents a harmonic oscillation that oscillates k times during one period T of $f(t)$. Thus, (C.1) indicates that an arbitrary periodic function with period T can be expressed as a combination of harmonic oscillations with angular frequency that is an integer multiple of $2\pi/T$.

When m and n are positive integers, the following relations hold:

$$\int_{-\pi}^{\pi} \cos mx \, dx = \int_{-\pi}^{\pi} \sin mx \, dx = 0, \quad \int_{-\pi}^{\pi} \cos mx \sin nx \, dx = 0,$$
$$\int_{-\pi}^{\pi} \cos mx \cos nx \, dx = \int_{-\pi}^{\pi} \sin mx \sin nx \, dx = \pi \, \delta_{mn}, \quad (C.2)$$

where δ_{mn} is the Kronecker delta defined by

$$\delta_{mn} = \begin{cases} 1 & (m = n), \\ 0 & (m \neq n). \end{cases} \quad (C.3)$$

(C.2) is called the "orthogonality" of trigonometric functions. Multiplying $\cos \omega_k t$ or $\sin \omega_k t$ on both sides of (C.1), integrating with respect to t for one period T, and using change of variables $x = \left(\frac{2\pi}{T}\right)t$ and the orthogonality (C.2), we can see that the coefficients a_k and b_k in (C.1) must be given by

$$a_k = \frac{2}{T} \int_0^T f(t) \cos \omega_k t \, dt \qquad (k = 0, 1, 2, \ldots), \tag{C.4a}$$

$$b_k = \frac{2}{T} \int_0^T f(t) \sin \omega_k t \, dt \qquad (k = 1, 2, \ldots). \tag{C.4b}$$

These are called **the Fourier coefficients** of $f(t)$, and such an infinite series on the right side of (C.1) that has these a_k and b_k as coefficients is called **the Fourier series** of $f(t)$.

If we use the relation

$$\cos \theta = \frac{1}{2} \left(e^{i\theta} + e^{-i\theta} \right), \quad \sin \theta = \frac{1}{2i} \left(e^{i\theta} - e^{-i\theta} \right), \tag{C.5}$$

which are immediately obtained from **Euler's formula** $e^{i\theta} = \cos \theta + i \sin \theta$, (C.1) and (C.4) can also be expressed in a complex form as follows:

$$f(t) = \sum_{k=-\infty}^{\infty} c_k \, e^{i\omega_k t}, \quad c_k = \frac{1}{T} \int_0^T f(t) \, e^{-i\omega_k t} \, dt \quad (k = 0, \pm 1, \pm 2, \ldots). \tag{C.6}$$

For arbitrary $f(t)$ which is piecewise continuous in $0 \leq t \leq T$, it can be shown that

$$\frac{1}{T} \int_0^T |f(t)|^2 \, dt = \frac{1}{4} a_0^2 + \frac{1}{2} \sum_{k=1}^{\infty} \left(a_k^2 + b_k^2 \right) = \sum_{k=-\infty}^{\infty} |c_k|^2 \tag{C.7}$$

holds. This is called **Parseval's identity**. The quantity on the left side of (C.7) often means the energy density of the signal $f(t)$, that is, the average energy contained per unit time. Considering that the subscript k of the Fourier coefficients is an index that distinguishes frequencies, (C.7) represents the energy density of $f(t)$ in the actual t space as a sum of the energy of each frequency component. In this sense, $\frac{1}{2} \left(a_k^2 + b_k^2 \right)$ and $|c_k|^2$ are called **energy spectrum** or **spectral intensity**.

From (C.7), it can be seen that

$$a_k, b_k \to 0 \quad (k \to \infty), \quad c_k \to 0 \quad (|k| \to \infty), \tag{C.8}$$

for arbitrary piecewise continuous function $f(t)$. This is known as **Riemann–Lebesgue lemma**.

C.2   FOURIER TRANSFORM

By considering a non-periodic function as a "periodic function of period ∞," we can extend the above results for a periodic function to a non-periodic function. The details are left to other textbooks, and only the basic results are shown here.

If $f(t)$ is smooth in any finite interval and absolutely integrable, that is, $\int_{-\infty}^{\infty} |f(t)|\, dt < \infty$, then

$$f(t) = \frac{1}{2\pi} \int_{-\infty}^{\infty} \left(\int_{-\infty}^{\infty} f(t') e^{-i\omega t'}\, dt' \right) e^{i\omega t}\, d\omega \tag{C.9}$$

holds. This is called **Fourier's integral theorem**. If the **Fourier transform** $F(\omega)$ of $f(t)$ is introduced by

$$F(\omega) = \frac{1}{\sqrt{2\pi}} \int_{-\infty}^{\infty} f(t)\, e^{-i\omega t}\, dt, \tag{C.10}$$

(C.9) can be written as

$$f(t) = \frac{1}{\sqrt{2\pi}} \int_{-\infty}^{\infty} F(\omega)\, e^{i\omega t}\, d\omega. \tag{C.11}$$

Note the symmetry of the representation for $f(t)$ and $F(\omega)$. This can also be written in real form as follows:

$$f(t) = \int_{0}^{\infty} \{A(\omega) \cos \omega t + B(\omega) \sin \omega t\}\, d\omega, \tag{C.12}$$

$$A(\omega) = \frac{1}{\pi} \int_{-\infty}^{\infty} f(t) \cos \omega t\, dt, \quad B(\omega) = \frac{1}{\pi} \int_{-\infty}^{\infty} f(t) \sin \omega t\, dt. \tag{C.13}$$

A periodic function $f(t)$ of period T can be expressed as a superposition of harmonic oscillations with discrete frequency at a constant interval $\Delta \omega = 2\pi / T$ as shown in (C.1). On the other hand, to represent a non-periodic function (i.e., period ∞), all the continuous frequencies ω are required, as indicated by (C.11) and (C.12).

For Fourier transforms, the equation

$$\int_{-\infty}^{\infty} |f(t)|^2 dt = \int_{-\infty}^{\infty} |F(\omega)|^2\, d\omega \tag{C.14}$$

holds, corresponding to Parseval's identity (C.7) for the Fourier series, and is called **Plancherel's theorem**. This expresses the total energy (left side) of $f(t)$ as an integral (right side) in the frequency ω space, and gives the basis of the concept of frequency spectrum.

C.3   SOLUTION OF THE DIFFUSION EQUATION

In Chapter 2, we introduced the solution (2.25) of the initial value problem of the diffusion equation in relation to the solution of the initial value problem of the Burgers equation. Here, let's obtain the solution of the initial value problem of the diffusion equation

$$\frac{\partial v(x,t)}{\partial t} = \nu \frac{\partial^2 v(x,t)}{\partial x^2}, \quad v(x,0) = v_0(x), \quad (-\infty < x < \infty), \tag{C.15}$$

using the knowledge of Fourier transform.

First, we will summarize the basic properties of the Fourier transform required for that purpose. Let $F(k)$ and $G(k)$ be the Fourier transforms of the function $f(x)$ and $g(x)$, respectively. Then the following properties hold.

(1) $\mathcal{F}\left[\dfrac{d}{dx}f(x)\right] = ikF(k)$.

(2) If the **convolution** $(f * g)(x)$ of $f(x)$ and $g(x)$ is defined by
$$(f * g)(x) \equiv \int_{-\infty}^{\infty} f(x - x')g(x')dx', \text{ then } \mathcal{F}[(f * g)(x)] = \sqrt{2\pi}\,F(k)G(k).$$

(3) $\mathcal{F}\left[e^{-ax^2}\right] = \dfrac{1}{\sqrt{2a}}e^{-k^2/4a}$ $(a > 0)$.

Take the Fourier transform of (C.15) with respect to x. If we write the Fourier transform of $v(x, t)$ and $v_0(x)$ with respect to x as $V(k, t)$ and $V_0(k)$, respectively,

$$\frac{\partial V(k, t)}{\partial t} = -vk^2\,V(k, t), \quad V(k, 0) = V_0(k) \quad \longrightarrow \quad V(k, t) = V_0(k)\,e^{-vk^2t}. \qquad \text{(C.16)}$$

Considering the property (3) above, $G(k, t) := e^{-vk^2t}$ that appears here is the Fourier transform of $g(x, t) = e^{-(x^2/4vt)}/\sqrt{2vt}$. Therefore, the result shown in (C.16) means that the Fourier transform $V(k, t)$ of $v(x, t)$ is given as the product of the Fourier transform $V_0(k)$ of $v_0(x)$ and the Fourier transform $G(k, t)$ of $g(x, t)$. Using property (2) above, this means that $v(x, t)$ is a convolution of $v_0(x)$ and $g(x, t)$, hence we can obtain the following expression for the solution $v(x, t)$:

$$v(x, t) = \frac{1}{\sqrt{4v\pi t}} \int_{-\infty}^{\infty} v_0(x')\exp\left[-\frac{(x - x')^2}{4vt}\right]dx'. \qquad \text{(C.17)}$$

APPENDIX D

Derivation of Governing Equations for Water Waves

This appendix introduces the minimum knowledge on fluid mechanics behind the governing equations (3.27) for water waves for readers who have not studied fluid mechanics.

D.1 MASS CONSERVATION LAW

Let $\rho(\boldsymbol{x},t)$, $\boldsymbol{v}(\boldsymbol{x},t)$, and $p(\boldsymbol{x},t)$ be the density, velocity, and pressure of fluid at position $\boldsymbol{x}$ and time t, respectively. Let's consider the conservation of mass, focusing on a certain region V in the fluid. The increase in mass per unit time in V is given by

$$\frac{d}{dt} \iiint_V \rho(\boldsymbol{x},t)\,dV. \tag{D.1}$$

On the other hand, the amount of mass flowing into V through the surface S of V per unit time is given by

$$-\iint_S \rho \boldsymbol{v} \cdot \boldsymbol{n}\,dS, \tag{D.2}$$

where $\boldsymbol{n}$ is the outward unit normal to S. (D.1) and (D.2) must be equal unless the fluid suddenly disappears or is created from nothing. Therefore,

$$\frac{d}{dt} \iiint_V \rho(\boldsymbol{x},t)\,dV = -\iint \rho \boldsymbol{v} \cdot \boldsymbol{n}\,dS. \tag{D.3}$$

Here, using that V is a fixed region and does not depend on time, and converting the surface integral to a volume integral by the divergence theorem (A.5), we obtain

$$\iiint_V \left\{ \frac{\partial \rho}{\partial t} + \operatorname{div}(\rho \boldsymbol{v}) \right\} dV = 0. \tag{D.4}$$

Since the region of interest V is arbitrary,

$$\frac{\partial \rho}{\partial t} + \operatorname{div}(\rho \boldsymbol{v}) = 0 \tag{D.5}$$

must hold everywhere in the fluid. This is called the **continuity equation** in fluid mechanics and physically represents mass conservation.

D.2 EQUATION OF MOTION

Next, let us consider the conservation of momentum, that is, the equation of motion. As in the previous section, focus on an arbitrary region V in the fluid. The change in momentum in V per unit time is given by

$$\frac{d}{dt} \iiint_V \rho \boldsymbol{v} \, dV. \tag{D.6}$$

Newton's equation of motion $ma = F$ (mass $\times$ acceleration $=$ force) implies that the force gives the rate of increase of momentum. Therefore, one cause of the momentum increase (D.6) in V is the force acting on the fluid in V.

There are two types of forces acting on the fluid. One is the "body force" that acts directly on the mass (or volume) of the fluid, such as gravity, and the other is the "surface force" that the surrounding fluid exerts on the fluid in V through the surface S of V, such as pressure. As the body force, we normally consider the force $\boldsymbol{F}_g$ from gravity given by

$$\boldsymbol{F}_g = \iiint_V \rho \boldsymbol{g} \, dV, \tag{D.7}$$

where $\boldsymbol{g} = (0, 0, -g)$ is the gravitational acceleration vector pointing vertically downward. On the other hand, in a motion where the effect of viscosity is negligible such as in water waves, the surface force consists only of the contribution $\boldsymbol{F}_p$ from pressure p and is given by

$$\boldsymbol{F}_p = - \iint_S p \boldsymbol{n} \, dS. \tag{D.8}$$

In addition to the these forces, there is another factor that increases the momentum in V. It is the inflow of external fluid to V through S. Since V is a region fixed in the space, the fluid enters and exits as the fluid moves. The increment in momentum per unit time associated with this inflow and outflow is given by

$$- \iint_S (\rho \boldsymbol{v}) \, \boldsymbol{v} \cdot \boldsymbol{n} \, dS. \tag{D.9}$$

Combining all of these, the conservation law of momentum is expressed as

$$\frac{d}{dt} \iiint_V \rho \boldsymbol{v} \, dV = \iiint_V \rho \boldsymbol{g} \, dV - \iint_S p \boldsymbol{n} \, dS - \iint_S (\rho \boldsymbol{v}) \, \boldsymbol{v} \cdot \boldsymbol{n} \, dS. \tag{D.10}$$

Here, if we convert the surface integral to a volume integral using the divergence theorem and write the i component ($i = 1, 2, 3$) of the vector equation, it becomes

$$\iiint_V \left[\frac{\partial (\rho v_i)}{\partial t} + \frac{\partial (\rho v_i v_j)}{\partial x_j} \right] dV = \iiint_V \left(-\frac{\partial p}{\partial x_i} + \rho g_i \right) dV. \tag{D.11}$$

Rearranging the integrand on the left side by using the continuity equation (D.5) and considering that the region V is arbitrary, we obtain

$$\frac{\partial v_i}{\partial t} + v_j \frac{\partial v_i}{\partial x_j} = -\frac{1}{\rho} \frac{\partial p}{\partial x_i} + g_i, \tag{D.12}$$

or in vector form

$$\frac{\partial \boldsymbol{v}}{\partial t} + (\boldsymbol{v} \cdot \nabla)\boldsymbol{v} = -\frac{1}{\rho} \nabla p + \boldsymbol{g}, \tag{D.13}$$

which must hold at any point in the fluid. This is the equation of motion that governs the fluid motion when the effect of viscosity is not taken into account, and is called the **Euler equation**.[1] By using the identity

$$(\boldsymbol{v} \cdot \nabla)\boldsymbol{v} = \nabla\left(\frac{1}{2}v^2\right) - \boldsymbol{v} \times (\nabla \times \boldsymbol{v}), \tag{D.14}$$

$\boldsymbol{g} = -\nabla(gz)$, and the fact that ρ can be treated as a constant for water, (D.13) can also be written as

$$\frac{\partial \boldsymbol{v}}{\partial t} = -\nabla\left(\frac{p}{\rho} + \frac{1}{2}v^2 + gz\right) + \boldsymbol{v} \times \boldsymbol{\omega}, \tag{D.15}$$

where $\boldsymbol{\omega}$ is a vector defined by $\boldsymbol{\omega} = \nabla \times \boldsymbol{v}$ and is called **vorticity**.

D.3 LAGRANGIAN DERIVATIVE

Let $F(\boldsymbol{x}, t)$ be a field of some physical quantity possessed by the fluid such as density and pressure. We can know the temporal rate of change of F at a certain point $\boldsymbol{x}$ by the partial derivative $\partial F(\boldsymbol{x}, t)/\partial t$ with respect to t. However, if the velocity at the point $\boldsymbol{x}$ is not zero, different fluid particles pass through this point one after another, so the partial derivative with fixed position is not the rate of change that a same fluid particle feels.

Then how can the rate of change perceived by a same fluid particle be evaluated? Suppose that we focus on the fluid particle which is at $\boldsymbol{x}$ at time t. The value of the physical quantity that this fluid particle has now is $F(\boldsymbol{x}, t)$. When the velocity field is $\boldsymbol{v}(\boldsymbol{x}, t) = (u, v, w)$, the small displacement $\Delta \boldsymbol{x}$ of this fluid particle during a short time Δt is given by $\Delta \boldsymbol{x} = \boldsymbol{v}\Delta t = (u\Delta t, v\Delta t, w\Delta t)$. Therefore, the change in F felt by the same fluid particle ΔF is given by

$$\begin{aligned}
\Delta F &= F(\boldsymbol{x} + \boldsymbol{v}\Delta t, t + \Delta t) - F(\boldsymbol{x}, t) \\
&= F(x + u\Delta t, y + v\Delta t, z + w\Delta t, t + \Delta t) - F(x, y, z, t) \\
&= \frac{\partial F}{\partial x} u\Delta t + \frac{\partial F}{\partial y} v\Delta t + \frac{\partial F}{\partial z} w\Delta t + \frac{\partial F}{\partial t} \Delta t \\
&= \Delta t \left(\frac{\partial F}{\partial t} + u\frac{\partial F}{\partial x} + v\frac{\partial F}{\partial y} + w\frac{\partial F}{\partial z}\right), \tag{D.16}
\end{aligned}$$

---

[1]When the effect of viscosity is taken into account, a term $(\mu/\rho)\nabla^2 \boldsymbol{v}$ is added on the right side, where μ is a material constant called **viscosity coefficient**. The equation with this viscosity term added to the Euler equation is called the **Navier–Stokes equation** and is the most important equation in fluid mechanics.

therefore, the rate of change of F felt by the same fluid particle is given by

$$\frac{\Delta F}{\Delta t} = \frac{\partial F}{\partial t} + u\frac{\partial F}{\partial x} + v\frac{\partial F}{\partial y} + w\frac{\partial F}{\partial z} = \left(\frac{\partial}{\partial t} + \boldsymbol{v} \cdot \nabla\right) F. \tag{D.17}$$

For an arbitrary physical quantity, when evaluating the rate of change felt by a fixed fluid particle, a differential operation consisting of a combination of temporal and spatial partial derivatives as

$$\frac{D}{Dt} = \frac{\partial}{\partial t} + (\boldsymbol{v} \cdot \nabla) \tag{D.18}$$

appears, and is called the **material derivative** or the **Lagrangian derivative**. The first term of the Lagrangian derivative is due to the temporal nonstationarity of the field, and the second term is due to the spatial nonuniformity of the field.

Using the Lagrangian derivative, (D.5) can be written as

$$\frac{D\rho}{Dt} + \rho \operatorname{div} \boldsymbol{v} = 0. \tag{D.19}$$

In liquid such as water, it is usually allowed to assume that the density ρ of fluid particle does not change, that is, $\frac{D\rho}{Dt} = 0$ holds. Such a fluid is called an **incompressible fluid**. For the velocity field $\boldsymbol{v}(\boldsymbol{x}, t)$ of an incompressible fluid,

$$\operatorname{div} \boldsymbol{v} = 0 \tag{D.20}$$

always holds.

D.4   KELVIN'S CIRCULATION THEOREM

When $\boldsymbol{v}$ is the velocity field, C is any simple closed curve in the fluid domain, and $d\boldsymbol{r}$ is line element of C, the line integral

$$\Gamma = \oint_C \boldsymbol{v} \cdot d\boldsymbol{r} \tag{D.21}$$

is called the **circulation** along C. Using Stokes' theorem in vector analysis, this can also be written as

$$\Gamma = \oint_C \boldsymbol{v} \cdot d\boldsymbol{r} = \iint_S (\nabla \times \boldsymbol{v}) \cdot \boldsymbol{n} \, dS, \tag{D.22}$$

where S is an arbitrary surface in the fluid region bounded by C, and $\boldsymbol{n}$ is its unit normal vector.[2]

When C is a material closed curve, that is, a closed curve that moves with the fluid particles, **Kelvin's circulation theorem**

$$\frac{D}{Dt} \oint_C \boldsymbol{v} \cdot d\boldsymbol{r} = 0 \tag{D.23}$$

---

[2]Here, the fluid region is assumed to be simply connected. Also, when using Stokes' theorem, the direction of $\boldsymbol{n}$ and the direction of going around C are linked to each other in the "right-hand screw relationship."

holds in the situation like fluid motion in water waves where the following three assumptions hold: (1) the fluid is inviscid, (2) external body force is conservative force like gravity, and (3) density ρ is constant. The circulation along a material closed curve is conserved if these assumptions hold. According to (D.22), (D.23) can also be written as

$$\frac{D}{Dt} \iint_S \boldsymbol{\omega} \cdot \boldsymbol{n} \, dS = 0. \tag{D.24}$$

(D.23) can be proved as follows.

$$\frac{D\Gamma}{Dt} = \frac{D}{Dt} \oint_C \boldsymbol{v} \cdot d\boldsymbol{r} = \oint_C \frac{D\boldsymbol{v}}{Dt} \cdot d\boldsymbol{r} + \oint_C \boldsymbol{v} \cdot \frac{D d\boldsymbol{r}}{Dt}. \tag{D.25}$$

If the velocity in the line element $d\boldsymbol{r}$ is the same everywhere, $d\boldsymbol{r}$ simply translates at this velocity and does not change in time. $d\boldsymbol{r}$ changes over time because of the difference in velocity. If $d\boldsymbol{v}$ is the difference in velocity at both ends of $d\boldsymbol{r}$, the change of $d\boldsymbol{r}$ per unit time, i.e., $D d\boldsymbol{r}/Dt$ is given by $d\boldsymbol{v}$ itself. Therefore, the second term on the rightmost side of (D.25) can be rewritten as

$$\oint_C \boldsymbol{v} \cdot \frac{D d\boldsymbol{r}}{Dt} = \oint_C \boldsymbol{v} \cdot d\boldsymbol{v} = \oint_C d\left(\frac{1}{2}v^2\right). \tag{D.26}$$

Considering that C is a closed curve and that $v^2 = |\boldsymbol{v}|^2$ is a single-valued function of space, this term is always 0.

Using the Lagrangian derivative $\frac{D}{Dt}$, and considering that ρ is constant and $\boldsymbol{g} = \nabla(-gz)$, the Euler equation (D.13) can be written as follows:

$$\frac{D\boldsymbol{v}}{Dt} = -\nabla\left(\frac{p}{\rho} + gz\right). \tag{D.27}$$

Inserting this, the first term of the rightmost side of (D.25) can be written as

$$\oint_C \frac{D\boldsymbol{v}}{Dt} \cdot d\boldsymbol{r} = -\oint_C \nabla\left(\frac{p}{\rho} + gz\right) \cdot d\boldsymbol{r} = -\left[\frac{p}{\rho} + gz\right]_{\text{start}}^{\text{end}}. \tag{D.28}$$

Considering that C is a closed curve, this is always 0, again. Thus, the circulation theorem (D.23) holds.

D.5   POTENTIAL FLOW AND BERNOULLI'S THEOREM

If the water is at rest initially, the circulation along any closed curve in the water region is of course zero initially. Then Kelvin's circulation theorem guarantees that the circulation along any closed curve in water region is zero at any later time. Considering (D.22), this means that the vorticity $\boldsymbol{\omega} = \nabla \times \boldsymbol{v}$ at any point in water is $\boldsymbol{0}$. In such a situation, according to "Helmholtz's decomposition theorem"[3] in vector analysis, the velocity field $v(x, t)$ can be expressed as $v = \nabla\phi$, that is, $v = (u, v, w) = (\phi_x, \phi_y, \phi_z)$, with an appropriate scalar function $\phi(x, t)$. Such ϕ is called the **velocity potential**. In this way, the flow whose vorticity is 0 everywhere in the fluid domain and the velocity vector can be expressed by a potential is called an **irrotational flow** or a **potential flow**.

When dealing with water waves, the change in density due to pressure is negligible, so (D.20) holds from the conservation law of mass. Substituting $v = \nabla\phi$ into this immediately gives the Laplace equation for ϕ

$$\nabla \cdot (\nabla\phi) = \frac{\partial^2\phi}{\partial x^2} + \frac{\partial^2\phi}{\partial y^2} + \frac{\partial^2\phi}{\partial z^2} = 0. \tag{D.29}$$

This is the field equation that the velocity potential $\phi(x, t)$ should satisfy at each point in the water region. When there is no y dependence, this equation gives (3.27a) in the system of governing equations for water waves.

Next, substituting $v = \nabla\phi$ into the Euler equation (D.15), and using $\omega = 0$, we can immediately obtain

$$\nabla\left(\frac{\partial\phi}{\partial t} + \frac{1}{2}v^2 + \frac{p}{\rho} + gz\right) = 0, \tag{D.30}$$

and by integrating this,

$$\frac{\partial\phi}{\partial t} + \frac{1}{2}\left[\left(\frac{\partial\phi}{\partial x}\right)^2 + \left(\frac{\partial\phi}{\partial y}\right)^2 + \left(\frac{\partial\phi}{\partial z}\right)^2\right] + \frac{p}{\rho} + gz = F(t), \tag{D.31}$$

where $F(t)$ is an arbitrary function that depends only on time. This is called the **(generalized) Bernoulli's theorem**. Consider this equation at the water surface. If the motion of air is ignored, the pressure at each point on the water surface is constant at atmospheric pressure p_0, and if there is no effect of surface tension, the water pressure is also equal to the atmospheric pressure p_0. In (D.31), choosing $F(t)$ equal to p_0, and assuming that there is no y dependence, (D.31) becomes

$$\frac{\partial\phi}{\partial t} + \frac{1}{2}\left[\left(\frac{\partial\phi}{\partial x}\right)^2 + \left(\frac{\partial\phi}{\partial z}\right)^2\right] + gz = 0, \tag{D.32}$$

at the water surface $z = \eta(x, t)$, which is equal to the dynamic boundary condition (3.27b) with the surface tension coefficient $\tau = 0$.

The effect of surface tension is directly linked to the curvature of the water surface. When the space is 3D and the surface profile is expressed as $z = \eta(x, y, t)$, we need to treat the curvature of surface, which would make the story somewhat complicated. So we assume here that

---

[3]According to Helmholtz's decomposition theorem, an arbitrary vector field $u(x)$ can be decomposed into the form $u = \nabla\phi + \nabla \times A$ using an appropriate scalar field $\phi(x)$ and a vector field $A(x)$. The first term is an irrotational vector field, and the second term is a solenoidal (i.e., divergent-free) vector field. In this case, $\phi(x)$ and $A(x)$ are called the scalar potential and the vector potential of $u(x)$, respectively.

there is no y-dependence in accordance with the treatment in the main text and that the water surface can be expressed as $z = \eta(x, t)$. Let us focus on the small line element ds on the water surface, as shown in Fig. D.1. The surface tension acts in the tangential direction with magnitude of $\tau[\mathrm{N}]$ (per unit length in the y direction) at both ends A and B of ds. If the water surface

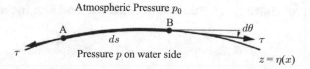

Figure D.1: Curvature of water surface.

is convex toward the air side as shown in the figure, a force toward the water side is generated as a resultant of the surface tension at A and B. To balance it, the pressure p on the water side becomes higher than the atmospheric pressure p_0, and a pressure jump Δp occurs between them. This Δp is given by

$$\Delta p \, ds = \tau \sin(d\theta) \approx \tau d\theta \quad \longrightarrow \quad \Delta p = \tau \frac{d\theta}{ds} = \tau\kappa, \qquad (D.33)$$

where $\kappa = \frac{d\theta}{ds}$ is the rate of change of the tangential direction per unit length that occurs when moving along the water surface and is called the **curvature**. The inverse of κ has a dimension of length and is called the **radius of curvature**. When the curve is represented as $z = \eta(x)$, the curvature κ can be evaluated by

$$\kappa = -\frac{\eta_{xx}}{(1 + \eta_x^2)^{3/2}}. \qquad (D.34)$$

Here, the curvature is signed so that it is positive when the water surface is convex to the air side, that is, the center of curvature is on the water side. When this extra pressure due to surface tension is added to (D.32), the dynamic boundary condition (3.27b) is obtained.

If there is a boundary of the fluid region due to the presence of an object, etc., the fluid cannot flow penetrating it. When the moving speed of the boundary is v_w, the normal component of velocity of the fluid must be equal to that of the boundary, i.e., $v \cdot n = v_w \cdot n$ must hold. This means that at the boundary the fluid moves only in the direction along the boundary. In the case of water waves, the water surface itself is a boundary of the fluid region, and the fluid on the surface moves only along the surface, therefore remains to exist on the surface at the next moment. Assume that there is no y dependency and the surface waveform is given by $z = \eta(x, t)$. If we introduce a scalar function $F(x, z, t)$ by $F(x, z, t) = \eta(x, t) - z$, $F(x, z, t)$ will always be 0 if we keep looking at the fluid particle on the surface, so $DF/Dt = 0$. This results in

$$\frac{DF}{Dt} = \frac{\partial F}{\partial t} + u\frac{\partial F}{\partial x} + w\frac{\partial F}{\partial z} = 0 \quad \longrightarrow \quad \frac{\partial \eta}{\partial t} + \frac{\partial \phi}{\partial x}\frac{\partial \eta}{\partial x} - \frac{\partial \phi}{\partial z} = 0, \qquad (D.35)$$

which leads to the kinematic boundary condition (3.27c). Similarly, the boundary condition (3.27d) corresponds to the fact that the vertical component ϕ_z of flow velocity at the bottom is 0 and the water does not penetrate the bottom.

As described above, all the equations and the boundary conditions of the system of governing equations for water waves used in Chapter 3 are derived. Refer to standard textbooks, such as [1, 3] and [2], for more systematic and detailed treatment of hydrodynamics and water waves.

D.6   REFERENCES

[1] G. D. Crapper. *Introduction to Water Waves*. Ellis Horwood, 1984. DOI: 10.1007/978-3-642-85567-2_15 206

[2] R. S. Johnson. *A Modern Introduction to the Mathematical Theory of Water Waves*. Cambridge U.P., 1997. DOI: 10.1017/cbo9780511624056 206

[3] P. K. Kundu and I. M. Cohen. *Fluid Mechanics*, 3rd ed., Elsevier, 2004. 206

APPENDIX E

Summary to Dimensional Analysis

In Chapter 3, it was shown that the dispersion relation of water waves can be derived (except for a dimensionless multiplicative coefficient) by simple consideration based on dimensions. Also, in the derivation process of important nonlinear wave equations such as the KdV equation and the NLS equation, the non-dimensionalization of the system of basic equations and rough estimate of the magnitude of each term in the equation played a very important role. In this appendix, we summarize the basic knowledge about the dimensions of physical quantities and the method of analysis based on them.

E.1 DIMENSION AND SI SYSTEM

In the **International System of Units (SI),** which is the most standard unit system in the world, the seven quantities such as length, mass, time, electric current, thermodynamic temperature, amount of substance, and luminous intensity are treated as the most basic quantities that appear in the laws of physics. If a quantity a represents a length, a is said to have the dimension of length, and we write $[a] = L$. Similarly, if the dimension of a is mass and time, we write $[a] = M$, $[a] = T$, respectively.[1] For physical quantities other than the basic quantities, we can know their dimensions by considering with what kind of calculations they can be derived from the basic quantities. For example:

- Density ρ: Density = Mass/Volume, [Volume] = $L \times L \times L$. Therefore, $[\rho] = ML^{-3}$;

- Velocity v: Velocity = Distance/Time. Therefore $[v] = LT^{-1}$;

- Force F: From Newton's law of motion $F = ma$, i.e., Force = Mass × Acceleration. Acceleration = Change of speed/Time. Therefore $[a] = LT^{-2}$. From this, $[F] = MLT^{-2}$; and

- Pressure p: Since pressure is a force that works per unit area, Pressure = Force/Area. Therefore, $[p] = [F]L^{-2} = ML^{-1}T^{-2}$, and so on.

---

[1]J. C. Maxwell, who established the fundamental system of equations of electromagnetics, is said to be the first to express the dimension of a by $[a]$.

E.2   PHYSICAL QUANTITIES WITH INDEPENDENT DIMENSIONS

Suppose there is a set of k physical quantities $\{a_1, a_2, \ldots, a_k\}$. For any one of them, if the multiplication and division of the other $(k-1)$ quantities cannot produce a quantity of the same dimension, the dimension of $\{a_1, a_2, \ldots, a_k\}$ is said to be independent, otherwise the dimension is dependent. In other words, the dimension of the set $\{a_1, a_2, \ldots, a_k\}$ is independent if the equation

$$[a_1]^{x_1}[a_2]^{x_2}\cdots[a_k]^{x_k} = 1, \tag{E.1}$$

has no non-trivial solution other than the trivial solution $\{x_1, x_2, \ldots, x_k\} = \{0, 0, \ldots, 0\}$, and if the equation has non-trivial solutions, the dimension is dependent.

As an example, let's examine whether the dimensions of density ρ, velocity v, and force F are independent or not:

$$[\rho] = ML^{-3}, \quad [v] = LT^{-1}, \quad [F] = MLT^{-2}. \tag{E.2}$$

$$[\rho]^x[v]^y[F]^z = 1 \longrightarrow \begin{cases} M: x + z = 0 \\ L: -3x + y + z = 0 \\ T: -y - 2z = 0 \end{cases} \longrightarrow \begin{pmatrix} 1 & 0 & 1 \\ -3 & 1 & 1 \\ 0 & -1 & -2 \end{pmatrix}\begin{pmatrix} x \\ y \\ z \end{pmatrix} = \begin{pmatrix} 0 \\ 0 \\ 0 \end{pmatrix}. \tag{E.3}$$

This system of linear equations for x, y, z has no solution other than the trivial solution $x = y = z = 0$, so the dimension of $\{\rho, v, F\}$ are independent. This means, for example, no matter how we multiply or divide ρ and v, we cannot produce a quantity with the same dimension as F.

Then, what about density ρ, velocity v, and pressure p?

$$[\rho] = ML^{-3}, \quad [v] = LT^{-1}, \quad [P] = ML^{-1}T^{-2}. \tag{E.4}$$

$$[\rho]^x[v]^y[p]^z = 1 \longrightarrow \begin{cases} M: x + z = 0 \\ L: -3x + y - z = 0 \\ T: -y - 2z = 0 \end{cases} \longrightarrow \begin{pmatrix} 1 & 0 & 1 \\ -3 & 1 & -1 \\ 0 & -1 & -2 \end{pmatrix}\begin{pmatrix} x \\ y \\ z \end{pmatrix} = \begin{pmatrix} 0 \\ 0 \\ 0 \end{pmatrix}. \tag{E.5}$$

In this case, there is a non-trivial solution $(x, y, z) = (-\alpha, -2\alpha, \alpha)$ (α is arbitrary). This solution means that ρv^2 has the same dimension as p.

As shown by the example above, whether or not the dimension of a set of physical quantities is independent is determined by whether or not the last homogeneous system of linear equations has a non-trivial solution. In the above two examples, the coefficient matrix becomes a 3×3 square matrix. The reason why the number of columns became 3 is because the target is a set of three physical quantities $\{\rho, v, F\}$ or $\{\rho, v, P\}$, while the reason why the number of raws became 3 is because three basic dimensions M, L, T are involved. For example, the first column of the coefficient matrix is the exponent of each basic dimension when the first quantity ρ is expressed by multiplication of powers of M, L, and T.

The dimension of a purely dynamical quantity that is not related to heat or electromagnetic phenomena is composed of three basic dimensions M, L, and T. If we do the same thing as (E.3) or (E.5) for such a set of four or more purely dynamical quantities, the number of columns of the coefficient matrix becomes larger that the number of raws, and non-trivial solution will always exist. Therefore a set consisting of four or more purely dynamical quantities is necessarily dimensionally dependent, and a quantity of the same dimension as one of the physical quantities can be created by multiplying or dividing other physical quantities.

E.3   CONVERSION OF UNIT SYSTEM

By determining one unit of the basic quantities (length, mass, etc.), the magnitude of any physical quantity can be expressed only by a single number that represents how many times the unit quantity it is. In the International System of Units (SI), length is measured in meter (m), mass is in kilogram (kg), time is in second (s), electric current in ampere (A), thermodynamic temperature in Kelvin (K), amount of substance in mole (mol), and luminous intensity in candela (cd) as one unit.

When converting to a new unit system in which 1 unit of mass is $1/\mathcal{M}$, 1 unit of length is $1/\mathcal{L}$, 1 unit of time is $1/\mathcal{T}$, etc. of the original unit system, the numerical value representing a physical quantity of dimension

$$[a] = M^\alpha L^\beta T^\gamma \tag{E.6}$$

becomes $\mathcal{M}^\alpha \mathcal{L}^\beta \mathcal{T}^\gamma$ times the numerical value in the original unit system. For example, the dimension of force F is MLT^{-2} as seen above. If we say $F = 3$ in the SI system, it is $3\,\mathrm{N}$, that is, $3\,\mathrm{kg\,m/s^2}$. When this is converted to the CGS unit system that uses gram, cm, and seconds as units of mass, length, and time, respectively, $\mathcal{M} = 1000$, $\mathcal{L} = 100$, $\mathcal{T} = 1$, therefore the numerical value of F becomes $F = 3 \times 10^5\,\mathrm{dyne}$, i.e., $3 \times 10^5\,\mathrm{g\,cm/s^2}$.

The following fact is very important for the future discussion.

[Theorem 1]

Let $\{a_1, a_2, \ldots, a_k\}$ be a set of k physical quantities with independent dimensions, and let A be an arbitrary real number. Then, there exists a method of conversion of the unit system (i.e., a new way to determine the one unit for each dimension) that changes any one of $\{a_1, a_2, \ldots, a_k\}$ to A times while keeps all other values unchanged.

Here we omit the proof,[2] and instead find actually the conversion of unit system so that only the numerical value of ρ is multiplied by A for the set $\{\rho, v, F\}$ whose dimensional independence was confirmed above. If the new unit amount of mass, length, and time are $1/\mathcal{M}$, $1/\mathcal{L}$, $1/\mathcal{T}$ of those in the current unit system, respectively, the numerical values representing $\{\rho, v, F\}$

---

[2]This can be proved if you have basic knowledge of linear algebra, so try it yourself.

change to $\mathcal{ML}^{-3}$ times, $\mathcal{LT}^{-1}$ times, and $\mathcal{MLT}^{-2}$ times, respectively. Therefore, it suffices to find $\mathcal{M}, \mathcal{L}, \mathcal{T}$ that satisfy

$$\mathcal{ML}^{-3} = A, \quad \mathcal{LT}^{-1} = 1, \quad \mathcal{MLT}^{-2} = 1. \tag{E.7}$$

Taking logarithms of these and setting $X = \log \mathcal{M}, Y = \log \mathcal{L}, Z = \log \mathcal{T}$, we obtain the required solution as follows:

$$\begin{pmatrix} 1 & -3 & 0 \\ 0 & 1 & -1 \\ 1 & 1 & -2 \end{pmatrix} \begin{pmatrix} X \\ Y \\ Z \end{pmatrix} = \begin{pmatrix} \log A \\ 0 \\ 0 \end{pmatrix} \longrightarrow X = Y = Z = -\frac{1}{2} \log A$$

$$\longrightarrow \mathcal{M} = \mathcal{L} = \mathcal{T} = A^{-1/2}. \tag{E.8}$$

This means that if we want only the numerical value representing ρ to be multiplied by 100 times (i.e., $A = 100$) while the numerical values of v and F to be unchanged, it can be achieved by converting to a new unit system in which the unit amount for measuring mass, length, and time are all 10 times the current unit system. Note that the coefficient matrix of the simultaneous linear equations that appears here is the transpose of the coefficient matrix of the homogeneous simultaneous linear equations when we judged the dimensional independence of $\{\rho, v, F\}$. Therefore, if the dimensions are independent, (E.8) always has a solution.

E.4   PI THEOREM

The purpose of many studies in science and engineering is to clarify the relationship between various physical quantities contained in the target phenomenon. As an example, consider the case of

$$a = f(a_1, a_2, a_3, b_1, b_2). \tag{E.9}$$

Here, a_1, a_2, a_3, b_1, b_2 represent the physical quantities contained in the phenomenon that can be known or can be controlled (control parameters), while a represents the objective physical quantity that we want to express as a function of the control parameters. We assume here that, of the control parameters, a_1, a_2, a_3 are physical quantities with independent dimensions, while b_1, b_2 are physical quantities whose dimensions can be expressed by a combination of dimensions of a_1, a_2, a_3 as

$$[b_1] = [a_1]^{p_1} [a_2]^{q_1} [a_3]^{r_1}, [b_2] = [a_1]^{p_2} [a_2]^{q_2} [a_3]^{r_2}. \tag{E.10}$$

In the first place, the natural world does not have a special length that should be used as a unit when measuring length. Using 1m as a unit length is only for human convenience. The same applies to the mass, time, and all the other quantities. Therefore, the laws of the natural world that hold regardless of the existence of human beings do not depend on the unit system. Newton's law of motion $F = ma$ (Force = mass × acceleration) always holds true whether we use SI unit or other unit system.

If a physical law like (E.9) holds, the dimension $[a]$ of the objective quantity a can always be expressed by multiplying dimensions of the dimensionally independent variables a_1, a_2, a_3 among the control parameters like

$$[a] = [a_1]^p [a_2]^q [a_3]^r. \tag{E.11}$$

Otherwise, $\{a, a_1, a_2, a_3\}$ has independent dimensions, but then from the above Theorem 1, we can arbitrarily change the value of a while keeping a_1, a_2, a_3 all unchanged by a suitable conversion of unit system, which means that a is not a function of a_1, a_2, or a_3.

Let's introduce the variables

$$\Pi = \frac{a}{a_1^p a_2^q a_3^r}, \quad \Pi_1 = \frac{b_1}{a_1^{p_1} a_2^{q_1} a_3^{r_1}}, \quad \Pi_2 = \frac{b_2}{a_1^{p_2} a_2^{q_2} a_3^{r_2}}, \tag{E.12}$$

in (E.9). Since the dimension of the denominator of Π, Π_1, Π_2 are the same as that of the numerator, they are all dimensionless quantities, and their values do not change if the unit system is changed. If we rewrite a, b_1, b_2 in (E.9) using these Π, Π_1, Π_2,

$$\Pi = \frac{1}{a_1^p a_2^q a_3^r} f\left(a_1, a_2, a_3, \Pi_1 a_1^{p_1} a_2^{q_1} a_3^{r_1}, \Pi_2 a_1^{p_2} a_2^{q_2} a_3^{r_2}\right). \tag{E.13}$$

The right side of this equation is a function of $a_1, a_2, a_3, \Pi_1, \Pi_2$, and if we write it as $\mathcal{F}(a_1, a_2, a_3, \Pi_1, \Pi_2)$, then

$$\Pi = \mathcal{F}(a_1, a_2, a_3, \Pi_1, \Pi_2). \tag{E.14}$$

In (E.14), a_1, a_2, a_3 have independent dimensions. Therefore, from Theorem 1, it is possible to change the value of a_1 to an arbitrary value while keeping a_2, a_3 unchanged by a conversion of the unit system. However, Π on the left side is a dimensionless quantity, and the value does not change even if this conversion of unit system is done. This means that Π does not depend on a_1. For the same reason, Π cannot depend on a_2 or a_3.

Therefore, the right side of (E.14) is a function only of Π_1 and Π_2, and it should be able to be written like

$$\Pi = \Phi(\Pi_1, \Pi_2) \tag{E.15}$$

with a certain suitable function Φ. If we return this to the relationship between the original quantities with dimensions, we obtain

$$a = a_1^p a_2^q a_3^r \, \Phi\left(\frac{b_1}{a_1^{p_1} a_2^{q_1} a_3^{r_1}}, \frac{b_2}{a_1^{p_2} a_2^{q_2} a_3^{r_2}}\right). \tag{E.16}$$

In this example, owing to the dimensional analysis, the original problem (E.9) which requires to deal with a function with five independent variables in order to express the relationship between the objective variable and the control parameters has been reduced to a much simpler problem (E.15) where we only need to deal with a function of two independent variables. The results so far are summarized in the following **Pi Theorem**.[3]

---

[3]It is said that E. Buckingham (1914) was the first to summarize the idea of dimensional analysis in the form of this Pi theorem. Incidentally, the name of the theorem, "Pi," has nothing to do with the circumference ratio π.

[Theorem 2: Pi Theorem]

The relation between an objective variable a and several control parameters a_1, $a_2, \ldots, a_k$, $b_1, \ldots, b_m$ can be rewritten as a relation between the dimensionless objective variable Π which is a made dimensionless by $a_1, a_2, \ldots, a_k$ and the dimensionless control variables $\Pi_1, \ldots, \Pi_m$ which are $b_1, \ldots, b_m$ made dimensionless by $a_1, a_2, \ldots, a_k$.

For example, in the case of (E.9), if we want to explore the relationship between the objective variable and the control parameters by a series of experiment or observation without doing any prior dimensional analysis at all, what will happen? We will probably do as follows. In order to find out the dependence of the objective variable a on the control parameter a_1, we first fix the values of the control parameters other than a_1, change only a_1 to various values and measure the value of a at each time, and plot the results on a graph paper or record it as a table. In order to see the trend, the value of a_1 will need to be changed to at least 10 different values or so. We need to repeat the same thing for the other four control parameters. Then at least 10^5 experiments or observations are required.

However, if the dimensional analysis is performed in advance and (E.9) has been reduced to the relationship (E.15), the experiments can be performed while changing Π_1 and Π_2, and we will need only 10^2 experiments.

In addition, there is other advantage as follows. The control parameters may contain physical quantities that are difficult to change their values, such as gravity g. It may also be difficult and costly to build an experimental apparatus that can change all of the control parameters. However, for (E.15), it is only necessary to change Π_1 and Π_2. We can do this by changing only the control parameters that are easy to change, without forcibly changing the control parameters whose values are difficult to change. This will make the procedure of experiments much simpler and the experimental apparatus much less expensive. In this way, thanks to the Pi theorem, the amount of experiments and observations (including direct numerical simulations) required to obtain the relationship between physical quantities can be greatly reduced, and the experiments themselves can be made much easier.

EXAMPLE 1: DIMENSIONAL ANALYSIS FOR PHASE VELOCITY OF SURFACE GRAVITY WAVES

The phase velocity c of the surface gravity wave should be expressed like

$$c = f(\rho, g, \lambda, h) \tag{E.17}$$

as a function of water density ρ, gravitational acceleration g, wavelength λ, and the water depth h. Consider this relationship by dimensional analysis using the Pi theorem.

[Answer]

The dimensions of the control parameters are $[\rho] = ML^{-3}$, $[g] = LT^{-2}$, $[\lambda] = L$, and $[h] = L$, respectively. Choose $\{\rho, g, \lambda\}$ as a set of control parameters with independent dimensions. Then h is dimensionally dependent. The dimension of the objective variable c is $[c] = LT^{-1}$, and the quantity with the same dimension as c in the product of $\{\rho, g, \lambda\}$ is $\sqrt{g\lambda}$. Therefore, the dimensionless relationship corresponding to (E.16) should be of the form[4]

$$c = \sqrt{g\lambda}\,\Phi(h/\lambda). \qquad (E.18)$$

This is all that we can do with dimensional analysis alone. After this, if we repeat experiments while changing the value of h/λ to determine the one-variable function $\Phi(x)$, then we can obtain information equivalent to the original relationship (E.17) which contains a function of four independent variables.

♣

. . . .

E.5 DRAG ON AN OBJECT BY DIMENSIONAL ANALYSIS

As an example of dimensional analysis, let us consider the drag experienced by a sphere moving in a fluid at a constant speed.

If we want to approach this problem seriously starting from the basic equation, we will need to solve the boundary value problem of the Navier–Stokes equation

$$\boldsymbol{v}_t + (\boldsymbol{v} \cdot \nabla)\boldsymbol{v} = -\frac{1}{\rho}\nabla p + \frac{\mu}{\rho}\nabla^2\boldsymbol{v}, \qquad (E.19)$$

which is the equation of motion of fluid dynamics. Especially when the velocity of the sphere is high, the flow around it becomes a very irregular and unsteady flow in a turbulent state, and it is totally impossible to solve this equation analytically. It is not easy to do numerical simulation either even using a state-of-the-art computer.

The dimensional analysis for this problem becomes as follows. The factors that may affect the drag F experienced by the sphere include the diameter d and the velocity U of the sphere, the density ρ and the viscosity coefficient μ of the fluid through which the sphere moves. Therefore, the goal is to find a relationship like

$$F = f(d, U, \rho, \mu). \qquad (E.20)$$

The dimensions of the relevant quantities are $[F] = MLT^{-2}$, $[d] = L$, $[U] = LT^{-1}$, and $[\rho] = ML^{-3}$. For the dimension of μ, some explanation may be necessary. In fluid dynamics, the force acting in a fluid is expressed by the quantity "stress." Stress τ is a force acting per unit

---

[4]As obtained in Chapter 3, the correct relationship is $c = \sqrt{g\lambda}\,\sqrt{\frac{1}{2\pi}\tanh\left(\frac{2\pi h}{\lambda}\right)}$.

area, so its dimension is the same as pressure and is given by $[\tau] = ML^{-1}T^{-2}$. Due to the property called "viscosity" of the fluid, if the flow velocity is not uniform, the fast fluid will drag the slow fluid and accelerate it, and conversely, the slow fluid will pull back the fast fluid and decelerate it. In ordinary fluids, the stress caused by this viscosity is considered to be directly proportional to the velocity gradient, and the proportionality factor is called the **viscosity coefficient** μ. Since the dimension of velocity gradient is $[v]/L = T^{-1}$, the dimension of μ is $[\mu] = [\text{stress}]/[\text{velocity gradient}] = ML^{-1}T^{-2}/T^{-1} = ML^{-1}T^{-1}$. The fact that μ has such a dimension can also be confirmed from the consistency of the dimensions of each term in (E.19).

In (E.20), $\{d, U, \rho\}$ of the control parameters have independent dimensions, and the dimension of μ can be expressed as $[\mu] = [d\,U\rho]$. Also, the dimension $[F]$ of the objective variable F can be expressed by $\{d, U, \rho\}$ as $[F] = [d^2 U^2 \rho]$.[5] Then, according to the Pi theorem, the dimensionless objective variable $\Pi = F/(d^2 U^2 \rho)$ is a function only of the dimensionless control parameter $\Pi_1 = \mu/(d\,U\rho)$, so F can be expressed as

$$F = d^2 U^2 \rho \times \Phi\left(\frac{\mu}{d\,U\rho}\right) \tag{E.21}$$

by an appropriate one-variable function Φ.

Incidentally, the reciprocal $\mathrm{Re} = \rho U d/\mu$ of the dimensionless control parameter Π_1 that appears here is called the **Reynolds number**, and is the most important dimensionless parameter in fluid dynamics. Reynolds number physically represents the ratio of inertial force to viscous force acting on the fluid, and the viscosity effect becomes more important as Re becomes smaller. In the same fluid, the larger the object and the faster the movement, the greater the Reynolds number.

As an example, let us estimate Re of the fastball of the baseball. Let the diameter be $d \approx 0.072$ (m) and the ball speed be 150 (km/h), i.e., $U \approx 42$ (m/s). For air, $\rho = 1.2$ (kg/m^3) and $\mu = 1.8 \times 10^{-5}$ (kg/m s). Then, $\mathrm{Re} \approx 2 \times 10^5$. This indicates that the effect of viscosity on the motion of the ball is about 10^{-5} times smaller than that of inertial force.

For motions where Re is sufficiently large, the effect of viscosity is considered to be negligible, and the drag on the sphere in such a situation may not depend on the viscosity coefficient μ. According to the dimensional analysis (E.21), in order for F to be independent of μ, Φ must be a constant function. If this is the case, at high Reynolds numbers, the drag F is expected to behave like

$$F = \alpha \rho d^2 U^2, \quad (\alpha : \text{dimensionless constant}) \tag{E.22}$$

implying that the drag is expected to be proportional to the square of the diameter of the sphere and the square of the speed of the sphere. This is known as **Newton's law of resistance**.

On the other hand, as an example of low Re, let us consider a case where a small plankton with a body length of 100 (μm) is swimming in water at a speed of 100 (μm/s). Then, $d = 1 \times 10^{-4}$ (m) and $U = 1 \times 10^{-4}$ (m/s). For water, $\rho = 1 \times 10^3$ (kg/m^3) and $\mu = 1.0 \times 10^{-3}$

---

[5] Speaking with (E.9), $\{d, U, \rho\}$ correspond to $\{a_1, a_2, a_3\}$, and μ corresponds to b_1. (b_2 does not exist.)

(kg/m s). Therefore, Re $\approx 10^{-2}$, that is, in this case, the effect of inertia is only 1/100 compared to the effect of viscosity. For the motion corresponding to such a small Reynolds number, the influence of fluid inertia is considered to be negligible compared to that of viscosity. The physical quantity that expresses inertia is mass, and it is ρ that reflects it in the present problem. (E.21) is not affected by inertia, that is, F does not include ρ only when $\Phi(x) = \beta x$, then (E.21) gives

$$F = \beta \mu U d. \quad (\beta : \text{dimensionless constant}) \tag{E.23}$$

This is know as the **Stokes' law of resistance**.[6]

Figure E.1 shows the relationship between the drag coefficient C_D defined by $C_D = F/\frac{1}{2}\rho U^2 (\pi d^2/4)$ and Re as a logarithm plot. The part where C_D is approximately proportional to $1/\text{Re}$ in the region of $\text{Re} < 1$ corresponds to Stokes' law of resistance, and the part around $10^3 < \text{Re} < 10^5$ where C_D is almost flat corresponds to the Newton's law of Resistance. Both Newton's law of resistance (E.22) at high Re and Stokes' law of resistance at low Re are very useful laws that give important information. What is used to obtain these laws is only dimensional considerations, and does not use the difficult governing equation (E.19) at all. The advantage of dimensional analysis is in the fact that it can be used when a phenomenon is too complex and we do not even know the equation that governs it.

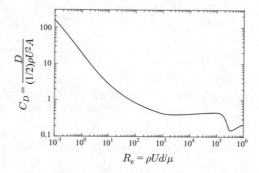

Figure E.1: Dependence of drag coefficient C_D of a sphere on Re.

For more information on the overall contents of this Appendix, see, for example, [1–3].

E.6   REFERENCES

[1] G. I. Barenblatt. *Scaling, Self-Similarity, and Intermediate Asymptotics: Dimensional Analysis and Intermediate Asymptotics.* Cambridge U.P., 1996. DOI: 10.1017/cbo9781107050242 215

---

[6]In 1909, Robert Millikan discovered in a series of experiments using charged oil droplets that the elementary charge (i.e., the charge of one electron) is about 1.6×10^{-19} coulomb. In this experiment, he used this Stokes' law of resistance in the essential part of the analysis of his experimental results. It seems very interesting that one of the most important basic quantities in the microscopic world has been found by using a law of very classical and macroscopic fluid mechanics.

[2] P. K. Kundu and I. M. Cohen. *Fluid Mechanics*, 3rd ed., Elsevier, 2004.

[3] A. C. Palmer. *Dimensional Analysis and Intelligent Experimentation*. World Scientific, 2008. DOI: 10.1142/6524 215

Derivation of the KdV Equation for Water Waves

In Chapter 5, the KdV equation was derived by an intuitive method of simply attaching the third derivative term u_{xxx} for expressing the dispersive character of the long-wavelength water waves to the nonlinear but non-dispersive wave equation derived by the "simple wave" theory explained in Appendix B. In this appendix, the KdV equation is derived more systematically by the perturbation method with the depth-to-wavelength ratio h/λ and amplitude-to-depth ratio a/h as small parameters. In the process of derivation, the long wave equation (B.3) will also be derived automatically.

F.1   THE BASIC EQUATIONS

Here, we show how to derive the long water wave equation (B.3) and the KdV equation (5.5) from the basic governing equations for surface water waves by a systematic method using perturbation expansion. The system of basic equations for surface water waves is given by

$$\phi_{xx} + \phi_{zz} = 0, \qquad -h \leq z \leq \eta(x,t) \qquad \text{(F.1a)}$$

$$\phi_t + gz + \frac{1}{2}\left(\phi_x^2 + \phi_z^2\right) = 0, \qquad z = \eta(x,t) \qquad \text{(F.1b)}$$

$$\eta_t + \phi_x \eta_x = \phi_z, \qquad z = \eta(x,t) \qquad \text{(F.1c)}$$

$$\phi_z = 0, \qquad z = -h, \qquad \text{(F.1d)}$$

as shown in Chapter 3.[1] The effect of surface tension is ignored here.

In the situation where the wavelength is very long compared to the water depth, there are various terms in the fundamental equations that are large and important or small and negligible. There is a method called **non-dimensionalization** as a conventional means used to know the rough relation between magnitude of each term as follows. First, for each variable, a "representative value" is introduced. For the horizontal coordinate x, the wavelength λ of the target wave, for the vertical coordinate z, the quiescent water depth h, for time t, the time required to propagate one wavelength λ at the typical propagation speed of long wave $\sqrt{gh}$, i.e., $\lambda/\sqrt{gh}$, and for η, the wave amplitude a are used as representative values, respectively. Considering that

[1]See Appendix D for their derivation.

the velocity of water particles is about $\sqrt{gh} \times \eta/h$, the reasonable choice for the representative value of the velocity potential ϕ will be $\sqrt{gh}\, a\lambda/h$.

Then introduce dimensionless variables by dividing each of the original variables by the corresponding representative values as follows:

$$\tilde{x} = x/\lambda, \quad \tilde{z} = z/h, \quad \tilde{t} = t/\left(\lambda/\sqrt{gh}\right) = \sqrt{gh}\, t/\lambda, \tag{F.2a}$$

$$\tilde{\eta} = \eta/a, \quad \tilde{\phi} = \phi/\left(\frac{\lambda a}{h}\sqrt{gh}\right). \tag{F.2b}$$

Since each variable is divided by its representative value, the magnitudes of all the dimensionless variables (shown with ˜) can be considered as $O(1)$.

Rewriting (F.1) using these dimensionless variables gives

$$\mu\, \tilde{\phi}_{\tilde{x}\tilde{x}} + \tilde{\phi}_{\tilde{z}\tilde{z}} = 0, \qquad\qquad -1 \leq \tilde{z} \leq \epsilon\tilde{\eta} \tag{F.3a}$$

$$\mu\left(\tilde{\phi}_{\tilde{t}} + \tilde{\eta}\right) + \frac{\epsilon}{2}\left\{\mu\left(\tilde{\phi}_{\tilde{x}}\right)^2 + \left(\tilde{\phi}_{\tilde{z}}\right)^2\right\} = 0, \qquad \tilde{z} = \epsilon\tilde{\eta} \tag{F.3b}$$

$$\mu\left(\tilde{\eta}_{\tilde{t}} + \epsilon\tilde{\phi}_{\tilde{x}}\tilde{\eta}_{\tilde{x}}\right) = \tilde{\phi}_{\tilde{z}}, \qquad\qquad \tilde{z} = \epsilon\tilde{\eta} \tag{F.3c}$$

$$\tilde{\phi}_{\tilde{z}} = 0, \qquad\qquad \tilde{z} = -1, \tag{F.3d}$$

where μ and ϵ are dimensionless parameters defined by

$$\mu = h^2/\lambda^2, \qquad \epsilon = a/h. \tag{F.4}$$

The situation of long wave (i.e., shallow water waves) whose wavelength is very long compared to the water depth corresponds to $\mu \ll 1$, while the situation where the amplitude of the wave is very small compared to the water depth and hence is close to linear corresponds to $\epsilon \ll 1$.

The most important point of this non-dimensionalization is that the quantities with ˜ are all $O(1)$ including their derivatives, so that all the magnitude relationships of terms in the equations are expressed explicitly in the coefficients. For example, in the long wave situation ($\mu \ll 1$), we can see at a glance that the first term on the left side of the Laplace equation (F.1a) is much smaller than the second term by looking at its dimensionless version (F.3a).

F.2   DERIVATION OF LONG WAVE EQUATION

Since there is no worry of confusion, we will omit the tilde (˜) indicating a dimensionless quantity. First, express ϕ in terms of a power series with respect to z around the bottom $z = -1$ as follows[2]:

$$\phi(x, z, t) = \sum_{n=0}^{\infty} (z + 1)^n\, \phi_n(x, t), \tag{F.5}$$

---

[2]Since the water depth is very shallow compared to the wavelength, it can be said that all the water region including the water surface is close to the bottom from the viewpoint of the wave. This leads to the idea of "expanding around the bottom."

where $\phi_1(x, t) = 0$ from (F.3d).

Differentiating (F.5) term by term gives

$$\phi_z = \sum_{n=1}^{\infty} n(z+1)^{n-1} \phi_n = \sum_{n=0}^{\infty} (n+1)(z+1)^n \phi_{n+1}, \tag{F.6a}$$

$$\phi_{zz} = \sum_{n=1}^{\infty} n(n+1)(z+1)^{n-1} \phi_{n+1} = \sum_{n=0}^{\infty} (n+2)(n+1)(z+1)^n \phi_{n+2}, \tag{F.6b}$$

$$\phi_x = \sum_{n=0}^{\infty} (z+1)^n \phi_{n,x}, \tag{F.6c}$$

$$\phi_{xx} = \sum_{n=0}^{\infty} (z+1)^n \phi_{n,xx}, \tag{F.6d}$$

where the subscripts x such as $\phi_{n,x}$ represents the partial derivative with respect to x.

Substituting (F.6b) and (F.6d) into (F.3a),

$$\mu \sum_{n=0}^{\infty} (z+1)^n \phi_{n,xx} + \sum_{n=0}^{\infty} (n+2)(n+1)(z+1)^n \phi_{n+2} = 0. \tag{F.7}$$

so

$$\phi_{n+2} = -\frac{\mu}{(n+2)(n+1)} \phi_{n,xx} \quad (n = 0, 1, \cdots). \tag{F.8}$$

Immediately from this and $\phi_1 = 0$, we obtain

$$\phi_3 = \phi_5 = \cdots = 0. \tag{F.9}$$

Equation (F.8) gives

$$\phi_2 = -\frac{\mu}{2} \phi_{0,xx}, \tag{F.10}$$

for $n = 0$ and

$$\phi_4 = -\frac{\mu}{4 \cdot 3} \phi_{2,xx} = \frac{\mu^2}{4!} \phi_{0,xxxx}, \tag{F.11}$$

for $n = 2$.[3] This gives

$$\phi = \phi_0 - \frac{\mu}{2}(z+1)^2 \phi_{0,xx} + \frac{\mu^2}{4!}(z+1)^4 \phi_{0,xxxx} + O(\mu^3), \tag{F.12}$$

as the expression of $\phi(x, z, t)$ which is correct up to $O(\mu^2)$.

---

[3]In general, $\phi_{2m} = O(\mu^m)$ from (F.8).

Substituting this result into the boundary conditions (F.3c) and (F.3b) yields

$$\eta_t + \epsilon\phi_{0,x}\eta_x + (1 + \epsilon\eta)\phi_{0,xx} = \frac{1}{2}\epsilon\mu(1 + \epsilon\eta)^2\phi_{0,xxx}\eta_x + \frac{1}{6}\mu(1 + \epsilon\eta)^3\phi_{0,xxxx} + O(\mu^2),$$
(F.13a)

$$\phi_{0,t} + \eta + \frac{\epsilon}{2}(\phi_{0,x})^2 = \frac{1}{2}\mu(1 + \epsilon\eta)^2\left\{\phi_{0,xxt} + \epsilon\phi_{0,x}\phi_{0,xxx} - \epsilon(\phi_{0,xx})^2\right\} + O(\mu^2). \quad \text{(F.13b)}$$

If we consider the long wave limit $\mu \to 0$ and ignore all terms including μ, we obtain

$$\eta_t + \epsilon\phi_{0,x}\eta_x + (1 + \epsilon\eta)\phi_{0,xx} = 0, \quad \text{(F.14a)}$$
$$\phi_{0,t} + \eta + \frac{\epsilon}{2}(\phi_{0,x})^2 = 0. \quad \text{(F.14b)}$$

Here, if we write $\phi_{0,x}(x,t)$ (i.e., the flow velocity at the bottom) as $u(x,t)$ and return to the original variables with dimensions, we obtain

$$\eta_t + u\eta_x + (h + \eta)u_x = 0, \quad \text{(F.15a)}$$
$$u_t + uu_x + g\eta_x = 0, \quad \text{(F.15b)}$$

which is the long wave equation (B.3). Thus, the derivation of the long wave equation assumes that the water depth is very shallow compared to the wavelength ($\mu = \sqrt{h/\lambda} \to 0$). However, no assumption is made about ϵ, so the long wave equation holds even if the situation is not close to linear. Such a framework with $\mu \to 0$, $\epsilon = O(1)$ is called the **Airy theory**. The dispersive character of water wave is not reflected in this theory.

F.3   DERIVATION OF THE KDV EQUATION

In (F.13a) and (F.13b), retaining terms with μ leads to inclusion of the effect of dispersion, and retaining terms with ϵ leads to inclusion of the effect of nonlinearity. It is difficult to capture both effects completely, so we aim at capturing only the minimum part of them. Specifically, considering μ and ϵ to be small parameters of the same order of magnitude, and taking only the first order terms with respect to them in (F.13a) and (F.13b), then we obtain

$$\eta_t + \epsilon\phi_{0,x}\eta_x + (1 + \epsilon\eta)\phi_{0,xx} = \frac{\mu}{6}\phi_{0,xxxx}, \quad \text{(F.16a)}$$
$$\phi_{0,t} + \eta + \frac{\epsilon}{2}(\phi_{0,x})^2 = \frac{\mu}{2}\phi_{0,xxt}. \quad \text{(F.16b)}$$

If we write the flow velocity $\phi_{0,x}(x,t)$ at the bottom as $u(x,t)$ as before,

$$\eta_t + \epsilon u\eta_x + (1 + \epsilon\eta)u_x - \frac{\mu}{6}u_{xxx} = 0, \quad \text{(F.17a)}$$
$$u_t + \eta_x + \epsilon uu_x - \frac{\mu}{2}u_{xxt} = 0. \quad \text{(F.17b)}$$

The theory that incorporates the lowest order effects of both dispersion and nonlinearity in this way is called the **Boussinesq theory**.

Since the z-dependence of the horizontal velocity is taken into account in the Boussinesq theory, the average flow velocity U is often used as the representative velocity instead of the flow velocity at the bottom. Differentiating (F.12) with respect to x,

$$\phi_x = \phi_{0,x} - \frac{\mu}{2}(z+1)^2\phi_{0,xxx} + O(\mu^2) = u - \frac{\mu}{2}(z+1)^2 u_{xx} + O(\mu^2). \qquad (F.18)$$

From this we find that the average velocity U is expressed in terms of u as

$$U = \frac{1}{1+\epsilon\eta}\int_{-1}^{\epsilon\eta}\phi_x\, dz = u - \frac{\mu}{6}(1+\epsilon\eta)^2 u_{xx} + O(\mu^2), \qquad (F.19)$$

from which u is given in terms of U as follows:

$$u = U + \frac{\mu}{6}(1+\epsilon\eta)^2 U_{xx} + O(\mu^2). \qquad (F.20)$$

Substituting this expression for u into (F.17a) and (F.17b) yields,

$$\eta_t + U_x + \epsilon U\eta_x + \epsilon\eta U_x = 0, \qquad (F.21a)$$
$$U_t + \eta_x + \epsilon U U_x - \frac{\mu}{3}U_{xxt} = 0, \qquad (F.21b)$$

which is rewritten as

$$\eta_t + hU_x + U\eta_x + \eta U_x = 0, \qquad (F.22a)$$
$$U_t + g\eta_x + U U_x - \frac{h^2}{3}U_{xxt} = 0, \qquad (F.22b)$$

in terms of variables with dimensions. This set of equations is called the **Boussinesq equation**.

By simplifying the Boussinesq equation by focusing on waves propagating in one direction, we can derive the KdV equations as follows. For the lowest-order approximation with $\mu = \epsilon = 0$, (F.21) becomes

$$\eta_t + U_x = 0, \quad U_t + \eta_x = 0. \qquad (F.23)$$

Differentiating the first and the second equations with respect to t and x, respectively, and taking the difference, we obtain

$$\eta_{tt} - \eta_{xx} = 0, \quad U_{tt} - U_{xx} = 0. \qquad (F.24)$$

Thus, in this approximation both $\eta(x,t)$ and $U(x,t)$ satisfy the well-known wave equation (F.24) and can be expressed as the sum of waves that translate left and right with the speed of ± 1 like

$$\eta(x,t) = f_+(x-t) + f_-(x+t), \qquad (F.25)$$

with the corresponding $U(x,t)$ which is linked to this $\eta(x,t)$ by (F.23), where f_+, f_- are arbitrary functions. Therefore, if we focus only on the wave traveling in the positive x direction, $\eta(x,t)$ and $U(x,t)$ satisfy

$$\eta_t + \eta_x = 0, \quad U_t + U_x = 0. \tag{F.26}$$

When compared with (F.23), this implies that, at the lowest-order approximation in μ and ϵ, both η and U translate in the positive x direction with speed 1 and that $\eta = U$.

The presence of terms of $O(\epsilon)$ and $O(\mu)$ in (F.21) introduces deviations from this simple translation. Because the cause of the deviation is the small terms of $O(\epsilon)$ and $O(\mu)$, the time scale for this deviation to occur is expected to be about $O(1/\epsilon)$. So, we introduce a new time variable $\tau = \epsilon t$ to treat this long time.[4] Since it is close to a translation with speed 1, we also introduce $\xi = x - t$ as a new space variable. Rewriting (F.21) by using the chain rule of partial differentiation,

$$\frac{\partial}{\partial t} = \frac{\partial}{\partial \xi}\frac{\partial \xi}{\partial t} + \frac{\partial}{\partial \tau}\frac{\partial \tau}{\partial t} = -\frac{\partial}{\partial \xi} + \epsilon\frac{\partial}{\partial \tau}, \quad \frac{\partial}{\partial x} = \frac{\partial}{\partial \xi}\frac{\partial \xi}{\partial x} + \frac{\partial}{\partial \tau}\frac{\partial \tau}{\partial x} = \frac{\partial}{\partial \xi}, \tag{F.27}$$

we obtain

$$-\eta_\xi + \epsilon\eta_\tau + U_\xi + \epsilon U\eta_\xi + \epsilon\eta U_\xi = 0, \tag{F.28a}$$

$$-U_\xi + \epsilon U_\tau + \eta_\xi + \epsilon U U_\xi + \frac{\mu}{3}U_{\xi\xi\xi} - \frac{\epsilon\mu}{3}U_{\xi\xi\tau} = 0. \tag{F.28b}$$

Here, ignoring $O(\epsilon\mu)$ and considering that the exchange of U and η is allowed in the terms of $O(\epsilon, \mu)$, we obtain by adding the two equations:

$$\epsilon\eta_\tau + \frac{3}{2}\epsilon\eta\eta_\xi + \frac{\mu}{6}\eta_{\xi\xi\xi} = 0. \tag{F.29}$$

Rewriting this in terms of x and t by

$$\frac{\partial}{\partial \tau} = \frac{1}{\epsilon}\left(\frac{\partial}{\partial t} + \frac{\partial}{\partial x}\right), \quad \frac{\partial}{\partial \xi} = \frac{\partial}{\partial x}, \tag{F.30}$$

we obtain the dimensionless form of the KdV equation:

$$\eta_t + \eta_x + \frac{3}{2}\epsilon\eta\eta_x + \frac{\mu}{6}\eta_{xxx} = 0. \tag{F.31}$$

Remembering that η, x, and t here are actually $\tilde{\eta}$, $\tilde{x}$, and $\tilde{t}$ introduced in (F.2), and substituting the definition (F.4) of ϵ and μ, we finally obtain the original KdV equation exactly the same as (5.7).

For more information on the subject of this Appendix, see, for example, [1, 2] and [3].

---

[4]Refer to Section 4.4 for more details on this idea of multiple scales.

F.4   REFERENCES

[1] R. S. Johnson. *A Modern Introduction to the Mathematical Theory of Water Waves*. Cambridge U.P., 1997. DOI: 10.1017/cbo9780511624056 222

[2] C. C. Mei. *The Applied Dynamics of Ocean Surface Waves*. John Wiley & Sons, 1983. DOI: 10.1142/9789812796059 222

[3] G. B. Whitham.   *Linear and Nonlinear Waves*.   John Wiley & Sons, 1974. DOI: 10.1002/9781118032954 222

[12] ... ...
... ...

[13] ... ...
... ...

[14] ... ...
... ...

APPENDIX G

FPU Recurrence and the KdV Equation

As mentioned in the footnote of Section 5.3.1, in the background of the numerical study of the KdV equation by Zabusky and Kruskal (1965), there was not only an interest as a water wave problem but also an awareness of a much bigger and fundamental problem related to the nonlinear science in general. In this appendix, we will introduce the motivation behind the study of Zabusky and Kruskal, while taking up the general properties of linear oscillation systems the unexpected phenomenon known as the "Fermi–Pasta–Ulam recurrence" appearing in numerical simulations of 1D nonlinear lattice systems, and the relationship between 1D nonlinear lattice systems and the KdV equation.

G.1 NORMAL MODE OF OSCILLATION

Let us consider the motion of a system in which two weights of mass m are connected by three identical springs with the spring constant k, as shown in Fig. G.1. It is assumed that the spring is

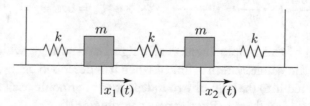

Figure G.1: A coupled system of mass and spring.

a linear spring for which the usual Hooke's law holds, and the potential energy when the spring is stretched by x is given by $\frac{1}{2}kx^2$. If the displacements of the weights from the equilibrium position at time t are $x_1(t)$ and $x_2(t)$, respectively, the kinetic energy T and the potential energy V are given, respectively, by

$$T = \frac{1}{2}m\dot{x}_1^2 + \frac{1}{2}m\dot{x}_2^2, \quad V = \frac{1}{2}kx_1^2 + \frac{1}{2}k(x_2 - x_1)^2 + \frac{1}{2}kx_2^2. \tag{G.1}$$

The Lagrangian L is defined by $L = T - V$, and corresponding Euler's equation of motion for x_1 and x_2 are given by

$$x_1: \quad \frac{d}{dt}\left(\frac{\partial L}{\partial \dot{x}_1}\right) - \frac{\partial L}{\partial x_1} = 0 \quad \longrightarrow \quad m\ddot{x}_1 = -2kx_1 + kx_2, \qquad \text{(G.2a)}$$

$$x_2: \quad \frac{d}{dt}\left(\frac{\partial L}{\partial \dot{x}_2}\right) - \frac{\partial L}{\partial x_2} = 0 \quad \longrightarrow \quad m\ddot{x}_2 = -kx_1 - 2kx_2. \qquad \text{(G.2b)}$$

Since x_1 and x_2 are not separated, we need to solve the simultaneous differential equations.

Although the reason is not explained in detail here, let us introduce the new variables ζ_1 and ζ_2 by

$$\zeta_1 = \sqrt{\frac{m}{2}}(x_1 + x_2), \qquad \zeta_2 = \sqrt{\frac{m}{2}}(x_1 - x_2). \qquad \text{(G.3)}$$

Then

$$x_1 = \frac{1}{\sqrt{2m}}(\zeta_1 + \zeta_2) \qquad x_2 = \frac{1}{\sqrt{2m}}(\zeta_1 - \zeta_2). \qquad \text{(G.4)}$$

Inserting these into (G.1), then T and V can be expressed as a sum of squares in terms of ζ_1 and ζ_2 as follows:

$$T = \frac{1}{2}\left(\dot{\zeta}_1^2 + \dot{\zeta}_2^2\right), \qquad V = \frac{1}{2}\left(\frac{k}{m}\zeta_1^2 + \frac{3k}{m}\zeta_2^2\right). \qquad \text{(G.5)}$$

As a result, Euler's equation of motion becomes

$$\zeta_1: \quad \frac{d}{dt}\left(\frac{\partial L}{\partial \dot{\zeta}_1}\right) - \frac{\partial L}{\partial \zeta_1} = 0 \quad \longrightarrow \quad \ddot{\zeta}_1 + \omega_1^2 \zeta_1 = 0, \quad \omega_1 = \sqrt{\frac{k}{m}}, \qquad \text{(G.6a)}$$

$$\zeta_2: \quad \frac{d}{dt}\left(\frac{\partial L}{\partial \dot{\zeta}_2}\right) - \frac{\partial L}{\partial \zeta_2} = 0 \quad \longrightarrow \quad \ddot{\zeta}_2 + \omega_2^2 \zeta_2 = 0. \quad \omega_2 = \sqrt{\frac{3k}{m}}. \qquad \text{(G.6b)}$$

Therefore, the motion of this coupled system of weights and springs, which looks like a complex motion in which the whole system is linked when it is described using intuitive coordinates (x_1, x_2), can be separated into the sum of two independent harmonic oscillations with frequencies ω_1 and ω_2 by describing them using the new coordinates (ζ_1, ζ_2).

The form of motion in which the whole system oscillates at a single frequency like (G.6) is called the **normal mode of oscillation** of the system, and the coordinates (ζ_1, ζ_2) used to separate into the normal modes is called the **normal coordinate**. In the case of the above example, for the normal mode represented by ζ_1 (i.e., $\zeta_2 = 0$), $x_1(t) = x_2(t)$, that is the displacement of weight 1 and weight 2 are always equal as shown in Fig. G.2a in which the central spring does not expand or contract. On the other hand, the normal mode represented by ζ_2 (i.e., $\zeta_1 = 0$), $x_1(t) = -x_2(t)$, and represents the form of motion in which the displacement of weight 1 and weight 2 are antisymmetric as shown in Fig. G.2b, and the central spring expands and contracts twice the springs on both sides, so that the restoring force against displacement is substantially tripled

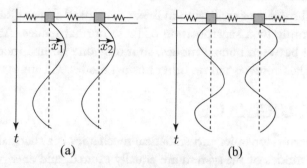

Figure G.2: Two normal modes of oscillation of the coupled system.

(hence $\omega_2 = \sqrt{3}\omega_1$). If the springs are linear obeying Hooke's law, the situation is essentially the same even when the number of weights is N, and there are N normal modes of oscillation with different frequencies and different mutual relationships of the displacement of each weight. And the general motion of the system is expressed by a linear combination of these normal modes of oscillation.

According to the general theory for small oscillation (linear theory) around the equilibrium point,[1] if x_i $(i = 1, \ldots, n)$ are the generalized coordinates representing the deviation from the equilibrium point, the kinetic energy T and the potential energy V are expressed as quadratic forms in $\dot{x}$ and x, respectively, as

$$T = \frac{1}{2}\sum_{i=1}^{n}\sum_{j=1}^{n} m_{ij}\dot{x}_i\dot{x}_j = \frac{1}{2}(\dot{x}, M\dot{x}), \quad V = \frac{1}{2}\sum_{i=1}^{n}\sum_{j=1}^{n} w_{ij}x_i x_j = \frac{1}{2}(x, Wx), \quad (G.7)$$

where $M = m_{ij}$ is a positive definite real symmetric matrix called **inertia matrix**, and $W = w_{ij}$ is a non-negative real symmetric matrix called **stiffness matrix**. Unless M and W are both diagonal, all of x_i will be interconnected in the equations of motion as in (G.2).

However, it is known that there is always a set of normal variables ζ_i $(i = 1, \ldots, n)$ obtained by an appropriate linear transform from x_i $(i = 1, \ldots, n)$ such that M and W can be diagonalized simultaneously when they are expressed in terms of ζ_i, and as a result, the Lagrangian can be expressed as a sum of squares of the normal variables like

$$L = T - V = \frac{1}{2}\sum_{i=1}^{n}\dot{\zeta}_i^2 - \frac{1}{2}\sum_{i=1}^{n}\omega_i^2\zeta_i^2. \quad (G.8)$$

Then, the set of Euler's equations of motion derived from this Lagrangian is separated into n independent equations containing only one coordinate ζ_i such as

$$\frac{d}{dt}\left(\frac{\partial L}{\partial \dot{\zeta}_i}\right) - \frac{\partial L}{\partial \zeta_i} = 0 \quad \longrightarrow \quad \ddot{\zeta}_i + \omega_i^2\zeta_i = 0 \quad (i = 1, \cdots, n). \quad (G.9)$$

[1]For more detail, see for example Chapter 6 of [3].

Thus, there are normal modes of oscillation in linear dynamical systems, and general motion of the system is given simply by a superposition of these normal modes. As indicated by (G.9), there is no interaction between normal modes, so if only one normal mode is excited initially, the normal mode will last forever, and no other normal modes will appear.

G.2   FPU RECURRENCE

The main object of thermodynamics and statistical mechanics is a thermal equilibrium state in which all the normal modes of the system are equally excited, and energy proportional to the absolute temperature is equally distributed to each normal mode oscillation. For example, it is an empirical fact that even if only a certain part of a material is heated or only a certain normal mode is excited initially, it will settle into such a thermal equilibrium state as time elapses. However, as described above, in the linear system, each normal mode behaves independently, so that no matter how much time passes, the system does not shift to a thermal equilibrium. Therefore, it is inferred that the driving force that allows the interaction (exchange of energy) between the normal modes and drives the entire system toward the thermal equilibrium should be the nonlinearity of the system.

For example, in the case of the coupled system with two weights connected with three springs as mentioned above, if the spring has a nonlinearity such that the potential energy is expressed by $\frac{1}{2}kx^2 - \alpha x^3$ instead of $\frac{1}{2}kx^2$, the Lagrangian L expressed in terms of the normal variables ζ_1, ζ_2 becomes

$$L = \frac{1}{2}\left(\dot{\zeta}_1^2 + \dot{\zeta}_2^2\right) - \frac{1}{2}\left(\omega_1^2\zeta_1^2 + \omega_2^2\zeta_2^2\right) + \frac{3\alpha}{\sqrt{2m^3}}(\zeta_1^2\zeta_2 - \zeta_2^3). \tag{G.10}$$

Therefore, the Euler equations of motion change from (G.6) to the following form:

$$\ddot{\zeta}_1 + \omega_1^2\zeta_1 = \frac{6\alpha}{\sqrt{2m^3}}\,\zeta_1\zeta_2, \qquad \ddot{\zeta}_2 + \omega_2^2\zeta_1 = \frac{3\alpha}{\sqrt{2m^3}}(\zeta_1^2 - 3\zeta_2^2), \tag{G.11}$$

and interactions between normal modes occur.

With this sort of awareness of problems, Fermi et al. (1955) [2] performed a series of numerical simulations of a 1D nonlinear lattice using an electronic computer called MANIAC that was just created. They considered a system in which up to 64 point masses are connected by a nonlinear spring that has a linear restoring force proportional to the first power of expansion (Hooke's law) as well as a restoring force proportional to the square or the third power of expansion, and traced the temporal evolution of the system numerically. At the initial time, they gave energy only to the normal mode that has the lowest frequency (the mode no.1 in Fig. G.3).

Figure G.3 shows the results of a numerical simulation similar to what they did. The horizontal and the vertical axises represent the time and the energy of each normal mode, respectively. In the early stage of evolution, it can be seen that, as they expected, normal modes

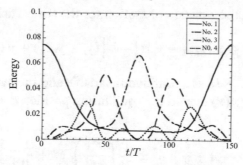

Figure G.3: FPU recurrence phenomenon.

other than no.1 are also excited due the interaction between the normal modes. However, contrary to their expectation that, after some sufficiently long time, the system would reach a thermal equilibrium in which all the normal modes were equally excited, the normal modes which are involved in energy redistribution process are limited to a relatively small number of modes with lower frequencies. Moreover, after a certain period of time, they observed a completely unexpected phenomenon to occur that almost all energy returned to the mode No.1 that was excited initially. This phenomenon is called **Fermi–Pasta–Ulam recurrence**, or FPU recurrence for short.[2]

In the course of research for the elucidation of this unexpected phenomenon that thermal equilibration does not necessarily occur even if the system has nonlinearity, Zabusky and Kruskal, who focused on the correspondence between the nonlinear lattice and the KdV equation as shown below, undertook the numerical study of the KdV equation.

G.3 DERIVATION OF THE KDV EQUATION FOR NONLINEAR LATTICE

The normal modes that are mainly excited in FPU recurrence phenomenon are those modes with a small mode number, that is, an oscillation with a long wavelength compared to the lattice spacing δ. Focusing on these modes, the original equation of motion, which is a set of n ordinary differential equations, can be approximated by one partial differential equation as shown below.

Let $y_i(t)$ be the deviation from the equilibrium position of the ith particle at time t. We assume that the spring has a weak nonlinearity and that the restoring force F is given by $F = -k(y + \gamma y^2)$, where k is the spring constant. The effect of nonlinearity is weak, so we

---

[2]The authors of the paper reporting the results of this study were three researchers, Fermi, Pasta, and Ulam. However, the realization of this research seems to have been greatly contributed by Mary Tsingou, a female researcher who developed the numerical algorithm and performed numerical simulations using an early-time computer. For this reason, in recent years it is sometimes called the "Fermi–Pasta–Ulam–Tsingou problem" with adding her name [1].

assume that $\gamma y \ll 1$. Then the equation of motion for the ith particle is given by

$$m \ddot{y}_i = k \left[(y_{i+1} - y_i) + \gamma (y_{i+1} - y_i)^2 \right] - k \left[(y_i - y_{i-1}) + \gamma (y_i - y_{i-1})^2 \right]. \qquad \text{(G.12)}$$

Regarding the coordinate $x = i\delta$ as a continuous variable and the displacement y as a function of x, and Taylor expanding $y(x)$ around the equilibrium position $x = i\delta$ of the ith particle, we obtain

$$y_{i\pm 1} = y_i \pm \left(\frac{\partial y}{\partial x} \right)_i \delta + \frac{1}{2} \left(\frac{\partial^2 y}{\partial x^2} \right)_i \delta^2 \pm \frac{1}{3!} \left(\frac{\partial^3 y}{\partial x^3} \right)_i \delta^3 + \frac{1}{4!} \left(\frac{\partial^4 y}{\partial x^4} \right)_i \delta^4 + \cdots. \qquad \text{(G.13)}$$

Inserting (G.13) into (G.12) yields

$$
\begin{aligned}
y_{tt} &= \frac{k}{m} \left[\left(y_x \delta + \frac{1}{2} y_{xx} \delta^2 + \frac{1}{3!} y_{xxx} \delta^3 + \frac{1}{4!} y_{xxxx} \delta^4 + \cdots \right) + \gamma \left(y_x \delta + \frac{1}{2} y_{xx} \delta^2 + \cdots \right)^2 \right] \\
&\quad - \frac{k}{m} \left[\left(y_x \delta - \frac{1}{2} y_{xx} \delta^2 + \frac{1}{3!} y_{xxx} \delta^3 - \frac{1}{4!} y_{xxxx} \delta^4 + \cdots \right) + \gamma \left(y_x \delta - \frac{1}{2} y_{xx} \delta^2 + \cdots \right)^2 \right] \\
&= \frac{k}{m} \left[\left(y_{xx} \delta^2 + \frac{1}{12} y_{xxxx} \delta^4 + \cdots \right) + 2\gamma y_x y_{xx} \delta^3 + \cdots \right], \qquad \text{(G.14)}
\end{aligned}
$$

and as an approximation we obtains

$$y_{tt} = c_0^2 \left(y_{xx} + 2\gamma \delta y_x y_{xx} + \frac{\delta^2}{12} y_{xxxx} \right), \qquad c_0^2 = \frac{k}{m} \delta^2. \qquad \text{(G.15)}$$

where c_0 gives the propagation velocity of the long-wave longitudinal wave traveling through this lattice system, when ignoring the effect of nonlinearity. If we employ the wavelength λ, the amplitude a, and λ/c_0 as the representative values of x, y, and t, respectively, the magnitude of each term on the right side of (G.15) relative to the left side is $O(1)$, $O((\gamma a)(\delta/\lambda))$, $O((\delta/\lambda)^2)$ in the order from the first term. The assumption of weak nonlinearity corresponds to $\gamma a \ll 1$, and the long-wave approximation corresponds to $\delta/\lambda \ll 1$. Assuming that $\gamma a \sim \delta/\lambda (= \epsilon)$, the last two terms on the right side are higher-order terms of $O(\epsilon^2)$.

Neglecting the higher order terms, (G.15) becomes the wave equation $y_{tt} = c_0^2 y_{xx}$, whose general solution is given by the **d'Alembert solution** $y(x,t) = f(x - c_0 t) + g(x + c_0 t)$, where f and g are arbitrary functions. By focusing only on the wave that propagates to the positive x direction, we can make (G.15) simpler. The solution of (G.15) translates at the speed of c_0 in the lowest order approximation, but is expected to deform slowly on a time scale of ϵ^{-2} due to the existence of the terms of $O(\epsilon^2)$. Reflecting this, we introduce new variables $\xi = x - c_0 t$ and $\tau = \epsilon^2 t$. Then,

$$\frac{\partial}{\partial t} = \frac{\partial}{\partial \xi} \frac{\partial \xi}{\partial t} + \frac{\partial}{\partial \tau} \frac{\partial \tau}{\partial t} = -c_0 \frac{\partial}{\partial \xi} + \epsilon^2 \frac{\partial}{\partial \tau}, \qquad \frac{\partial}{\partial x} = \frac{\partial}{\partial \xi} \frac{\partial \xi}{\partial x} + \frac{\partial}{\partial \tau} \frac{\partial \tau}{\partial x} = \frac{\partial}{\partial \xi}. \qquad \text{(G.16)}$$

Inserting this to (G.15) gives

$$\epsilon^4 y_{\tau\tau} - 2\epsilon^2 c_0 y_{\tau\xi} + c_0^2 y_{\xi\xi} = c_0^2 \left(y_{\xi\xi} + 2\gamma\delta y_\xi y_{\xi\xi} + \frac{\delta^2}{12} y_{\xi\xi\xi\xi} \right). \tag{G.17}$$

If we neglect $O(\epsilon^4)$ here and rewrite y_ξ as u, we obtain

$$\epsilon^2 u_\tau + c_0\gamma\delta u u_\xi + \frac{c_0\delta^2}{24} u_{\xi\xi\xi} = 0, \tag{G.18}$$

which gives the following KdV equation when written in the original coordinates x, t,

$$u_t + c_0 u_x + \alpha u u_x + \beta u_{xxx} = 0, \qquad \alpha = c_0\gamma\delta, \quad \beta = \frac{c_0\delta^2}{24}. \tag{G.19}$$

In this way, the set of equations of motion (G.12) of the nonlinear lattice can be approximated by the KdV equation if the wavelength of the motion is long relative to the lattice spacing δ and the effect of nonlinearity is weak.

G.4   REFERENCES

[1] T. Dauxois. Fermi, Pasta, Ulam, and a mysterious lady. *Physics Today*, 61:55–57, 2008. DOI: 10.1063/1.2835154 229

[2] E. Fermi, J. Pasta, and U. Ulam. Studies of non linear problems. In *Collected Papers of Enrico Fermi*, vol. 2, 1965. DOI: 10.2172/4376203 228

[3] H. Goldstein. *Classical Mechanics*. Addison-Wesley, 1980. DOI: 10.2307/3610571 227

Author's Biography

MITSUHIRO TANAKA

Mitsuhiro Tanaka is a Professor Emeritus at Gifu University and also a visiting professor at The Open University of Japan. After receiving his Doctor of Science from Kyoto University in 1983, he has been engaged in education and research in applied mathematics and fluid mechanics at Gifu University for more than 30 years.

Index